CREATIVE ECONOMIES, CREATIVE COMMUNITIES

Creative Economies, Creative Communities

Rethinking Place, Policy and Practice

Edited by

SASKIA WARREN
University of Manchester, UK

and

PHIL JONES
University of Birmingham, UK

ASHGATE

Published by
Ashgate Publishing Limited
Wey Court East
Union Road
Farnham
Surrey, GU9 7PT
England

Ashgate Publishing Company
110 Cherry Street
Suite 3-1
Burlington, VT 05401-3818
USA

www.ashgate.com

British Library Cataloguing in Publication Data
A catalogue record for this book is available from the British Library

The Library of Congress has cataloged the printed edition as follows:
Creative economies, creative communities : rethinking place, policy and practice / [edited] by Saskia Warren and Phil Jones.
pages cm
Includes bibliographical references and index.
ISBN 978-1-4724-5137-8 (hardback) -- ISBN 978-1-4724-5138-5 (ebook) -- ISBN 978-1-4724-5139-2 (epub) 1. Cultural industries--Social aspects. 2. Cultural geography. 3. Cultural policy. 4. Community development. I. Warren, Saskia. II. Jones, Phil (Phil Ian)
HD9999.C9472C736 2015
306.3--dc23

2015012392

ISBN: 9781472451378 (hbk)
ISBN: 9781472451385 (ebk – PDF)
ISBN: 9781472451392 (ebk – ePUB)

MIX
Paper from responsible sources
FSC® C013985
FSC www.fsc.org

Printed in the United Kingdom by Henry Ling Limited, at the Dorset Press, Dorchester, DT1 1HD

Contents

List of Figures

List of Tables

Notes on Contributors

Tim Acott is a Lecturer in Environmental Geography at the University of Greenwich. He graduated with a BSc Hons in Environmental Science from the University of Plymouth in 1989. He subsequently completed a PhD at the University of Stirling and started to lecture at the University of Greenwich in 1993. Tim was a lead investigator on a European Interreg 4a funded collaborative project, CHARM III (Channel Integrated Approach for Marine Resource Management), a €4.6 million ERDF co-financed Interrg IVA 2 Seas project called Geography of Inshore Fishing and Sustainability (GIFS) and the Interreg IVA 2 Seas TourFish (Tourism for Inshore Fishing, Food and Sustainability). Tim has worked on social science research projects spanning marine fishing, environmental conservation, sustainable tourism, ecotourism and environmental ethics.

Roxanna Collins works within Birmingham City Council's Culture and Visitor Economy Service. Over the last nine years, she has enjoyed working with residents and partners to support the development of arts and culture through various community programmes and in a variety of Council roles, from policy to regeneration. Roxanna completed an MA in Community and Participatory Arts at Staffordshire University (2012), and a BA in Fine Art and History of Art and Architecture at the University of Reading (2005), where she developed an interest in town planning, Brutalist and post-war architecture, and art in the public realm. Roxanna is also a committee member of the West Midlands Twentieth Century Society, and community arts organisation, Stirchley Happenings, and advisor for Still Walking Festival.

Shari Daya is a Lecturer in Human Geography at the University of Cape Town. Her doctoral research explored discursive constructions of the 'new Indian woman', interrogating the gendered and ethnocentric nature of dominant narratives of modernity. After completing her PhD in 2007, she worked on a range of research projects, on topics including sustainable consumption, lay ethics and nanotechnologies, and the relationship between adolescent wellbeing and neighbourhoods. Her current research explores the values underlying production and consumption practices in South Africa, with a particular focus on 'alternative' economies, such as those characterised by informality, creativity or morality.

Jarosław Działek is a Lecturer and Researcher in Human Geography at the Department of Regional Development, Institute of Geography and Spatial Planning at the Jagiellonian University in Krakow. He holds an MA in Sociology and a PhD in Geography. His research focuses on the problems of local and regional development, especially its social dimensions, social capital and regional identity, as well as development of knowledge economy, innovation, cooperation between science, institutions and firms. He is an author of over 30 publications including a monograph *Social Capital as a Factor of Economic Development at the Regional and Local Level in Poland* (in Polish, 2011).

Steve Harding heads the Regional Team in Birmingham City University with a focus on cross innovation across growth sectors informed by BCU's leadership in the Interreg IVC project. He holds a PhD from the University of Nottingham (2002). He is a Knowledge Integrator with the University of Lancaster in the Design for Europe platform to support design-led innovation across the EU. Steve led the writing team for the region's European Structural and Investment Fund (ESIF) submission for 2014–20.

He is a member of the NICE network led by the European Centre for the Creative Economy in Essen promoting creative spillovers.

Phil Jones is a Senior Lecturer in Cultural Geography at the University of Birmingham and Principal Investigator on a £1.5m AHRC-funded project *Cultural Intermediation and the Creative Urban Economy* (2012–16). He has published widely on cities, including work on sustainability and regeneration as well as arts and creative methods. Along with James Evans, Phil is co-author of the book, *Urban Regeneration in the UK* (second edition 2013).

Antonia Layard is a Professor at Bristol Law School, University of Bristol. Antonia's research is in law and geography where she explores how law, legality and maps construct space, place and 'the local'. She has particular interests in the legal provisions and practices involved in large-scale regeneration and infrastructure projects, and teaches courses on property, planning and environmental law. *Law, Place and Maps*: *Balancing Protection and Exclusion* will be published by Glasshouse Press, Routledge. Until September 2014, she was an ESRC Fellow researching Localism, Law and Governance.

Paul Long is Director of the Birmingham Centre for Media and Cultural Research, Birmingham City University. He is the co-author of 'Voicing Passion: The Emotional Economy of Songwriting' in *European Journal of Cultural Studies* (2015) and 'Online Archival Practice and Virtual Sites of Musical Memory' in Les Roberts, Marion Leonard, Sara Cohen and Robert Knifton (eds), *Sites of Popular Music Heritage* (Routledge, forthcoming). He has provided research on the Interreg-funded project *Cross Innovation, Promoting Cross Innovation in European Cities and Regions* and is Co-Investigator on the AHRC-funded *Cultural Intermediation: Connecting Communities in the Creative Urban Economy*.

Jane Milling is Assistant Professor in Drama at the University of Exeter where she researches popular and political performance in British theatre history in the long eighteenth-century and the modern period. She is interested in questions of participation, community, and creativity in contemporary performance culture and is currently working on two AHRC-funded projects under the Connected Communities programme. She has commissioned and edited monograph series for Palgrave and Ashgate, and sits on the editorial board for the journals *Theatre, Dance and Performance Training,* and *RECTR*.

Monika Murzyn-Kupisz is Assistant Professor in the UNESCO Chair for Heritage and Urban Studies, Department of Economic and Social History, Krakow University of Economics, Poland. She is a doctor of Economic Sciences (Krakow University of Economics, Poland) and holds an MA in European Leisure Studies (a joint diploma of the Vrije Universiteit Brussel, Tilburg University, Universidad de Deusto in Bilbao and Loughborough University) and a post-graduate diploma in Heritage Management. In 2000–2009 she worked as a senior specialist at the Research Institute of European Heritage, International Cultural Centre in Krakow. She is also a member of ICOMOS Poland. Her research specialises on contemporary attitudes, usage and interpretation of heritage, cultural economics and cultural policy, as well as urban regeneration and contemporary transformations of historic cities, with a special focus on Central and Eastern Europe. She is an author of over 50 publications on the above subjects, including two monographs *Kazimierz: The Central European Experience of Urban Regeneration* (in English and Polish, 2006), and *Heritage and Local Development* (in Polish, 2012).

Dave O'Brien is a Senior Lecturer in cultural policy at Goldsmiths, University of London. A sociologist and political scientist, his work covers questions of urban regeneration, cultural policy and cultural value. His most recent book, *Cultural Policy*, was published by Routledge in 2013.

Liz Roberts is a Research Fellow working at the University of West of England. Her research has a focus on creative practices as well as exploring different ways in which communities can develop processes for building resilience. Her work at the RCUK dot.rural Digital Economy Hub in Aberdeen looked to the potential of new digital technologies as enablers for resilience in rural communities, whilst her current project engages with narrative/creative approaches to water scarcity and resource management. Liz gained her PhD in Human Geography from the University of Exeter. Her research interests include cultural and digital geographies, community-facing research for resilience and social justice, visual and new media studies, and creative methodologies.

Leanne Townsend is a researcher working across various projects at the dot. rural Digital Economy Hub at the University of Aberdeen. Leanne's current work explores: the impact of broadband (and lack of) for rural creative practitioners across Scotland and the South West of England; barriers to digital adoption of broadband in rural areas; the role of social media in developing social capital for rural communities and businesses; the impacts of digital technologies on identities of place. She is leading research concerned with digital narratives and engagement for rural craftspeople. Leanne gained her PhD in Environmental Psychology from Robert Gordon University, Aberdeen.

Julie Urquhart is a rural geographer and social scientist at the Centre for Environmental Policy, Imperial College London. She has a first class BSc (Hons) in Environmental Science, distinction at MA in Research Methods (for Countryside Planning) and completed her PhD in forest policy and economics in 2009 at the Countryside and Community Research Institute, University of Gloucestershire. Currently Julie is working on the UNPICK project (funded under the auspices of the Living With Environmental Change Partnership), which aims to investigate how UK publics understand and perceive the growing threats to tree health from invasive pests and diseases. Prior to this, she was co-investigator on a €4.6 million ERDF co-financed Interreg IVA project GIFS, the Geography of Inshore Fishing and Sustainability and the Interreg IVA 2 Seas TourFish (Tourism for Inshore Fishing, Food and Sustainability) cluster. Julie has also worked on the European Interreg 4a funded collaborative project, CHARM III (Channel Habitat Atlas for Resource Management) and has undertaken contract work for Defra, English Heritage and the Forestry Commission in areas such as the social impacts of marine fishing, economic evaluation of heritage in National Parks, and woodland management and public good outputs.

Saskia Warren is a Lecturer in Human Geography at the University of Manchester. She has recently worked on a cross-sectoral and interdisciplinary £1.5m AHRC-funded project, *Cultural Intermediation and the Creative Urban Economy* based at the University of Birmingham. Saskia led a Communities and Culture Network+ funded project, Birmingham Surrealist Laboratory, which investigates the ways in which new digital facilities can unlock complex issues of community and cultural heritage in a diverse city. She completed her PhD in Cultural Geography at Sheffield (2012) and MA in Art Gallery and Museum Studies at Leeds (2008). She has worked as a curator, arts consultant and policy-maker, and as Research and Exhibition Assistant, Faculty of Art, Design and Architecture, Kingston University (2011–12).

Ginnie Wollaston works within Birmingham City Council's Culture and Visitor Economy Service and was Dance Officer for Arts Council West Midlands from 2002–5. She has developed her interest in leading arts and communities work using cultural co-design through a city-wide programme called *Connecting*

Communities through Culture co-funded by DCLG, Arts Council England and Birmingham City Council. Previously she had a professional career in managing educational and outreach programmes for professional dance companies (Shobana Jeyasingh Dance Company and Extemporary Dance Theatre). Her academic work includes an MA in Dance and Anthropology (1999); Diploma in Dance and Dance Education, and a degree in Fine Art.

Chapter 1
Introduction

Saskia Warren and Phil Jones

Within western policy discourses the creative economy is often positioned as a magic bullet that will save cities from the ill effects of post-industrial decline (Landry and Bianchini, 1995; Florida, 2002). A barrage of statistics can be marshalled to demonstrate the added value of the creative sector to national economies (Bakhshi et al., 2013). While some sub-sectors such as music and fashion are presented as having low barriers to entry (Hracs et al., 2013), important recognition must be given to barriers of skill and social capital (Hracs, 2013), inequalities of entry according to gender, race, class and age (Donald et al., 2013), along with high educational barriers in certain sub-sectors, including museums, galleries and archival work, which serve to restrict entry and condition success. For instance, the Department for Culture Media and Sport found that over 40 per cent of UK creative workers have degrees in comparison to 16 per cent in other sectors (DCMS, 2006). For those trapped in low skill jobs in the service sector, there are still tremendous challenges in seeing the creative economy as offering an obvious route out of structural inequality and into career pathways.

Neoliberal policy assemblages work to reduce the power of the state, in particular its role in redistributing the spoils of economic growth. 'Communities' – ambiguous and contested – are increasingly positioned as the guardians of social justice, juxtaposed against a supposedly remote and arbitrary public sector (Imrie and Raco, 2003). This puts an increasing amount of responsibility on communities to mitigate the more deleterious effects of neoliberal capitalism as well as to oversee what were traditionally state-run public services. Communities have, for example, been asked to take an increasingly direct role in schools, hospitals, libraries, and even spatial planning (Allmendinger and Haughton, 2012).

When it comes to the creative sector, the idea of community is often positioned in terms of a deficit model – individuals lack the skills and social capital to reap the rewards of the creative economy and therefore communities become a vehicle and a target for activity to address these deficits. This can be crudely characterised as a desire to 'fix' communities so that individuals can meaningfully engage with the creative economy and gain from its benefits. It also suggests that the creative sector should act as an agent for tackling structural inequality, regardless of whether that is where the expertise of individuals within that sector lies (Oakley, 2009).

Discussions about creative labour have considered its relationship with the organisation of power and space. Rob Pope (2005: xvi) defines creativity as 'the capacity to make, do or become something fresh and valuable with respect to others as well as ourselves'. This capacity, however, is subject to socio-spatial contexts that shape access to the sector, division of labour and pay conditions. Angela McRobbie (2002: 526) distinguishes between perceptions of creative labour as inclusive with the reality of how the sector functions:

> There is an irony in that alongside the assumed openness of the [creative] network, the apparent embrace of non-hierarchical working practises, the various flows and fluidities … there are quite rigid closures and exclusions.

Issues of access are a key concern. While it remains important to be aware of the agendas driving social inclusion of disadvantaged communities into the creative economy, this edited collection starts from the position that if public funding is being directed towards growth in the creative and cultural sector then it is necessary to investigate who is benefiting. Barriers to gaining regularly paid employment are indicative that claims for the democratising tendencies and pathways for the socially excluded in the creative sector have been exaggerated. The work of cultural intermediaries – those who act as mediators, brokers and gatekeepers – cannot necessarily be considered innovative, ground-breaking or, indeed, democratising (see Negus, 2002). Furthermore, there is a significant tension between economic enterprise and growth objectives, aesthetic excellence and non-monetary exchange values of social inclusion for disadvantaged communities.

This collection explores the intersection of creative economy and notions of community, examining connections and disconnections within policy and practice. The focus is primarily within Europe including a post-socialist example, although it also draws upon a case study from the developing world to examine how these discourses are being deployed and realised in a wider context. At its heart, this book brings the idea of *place* to the fore, exploring how specific localities fundamentally shape the connections between creative economy and communities. In this introduction we review the conceptual frame underpinning the arguments within this book. First, we contextualise the policy debates that have helped to shape the contemporary creative economy. Secondly, we examine how communities and creative economy practice have been drawn together within this policy framing and by the academy. Finally we consider how the idea of place can be used to explore how policy, community and practice co-construct each other within specific local contexts. We then outline the key contributions of each chapter within this book and detail the overarching themes that span the collection.

Policymaking, the Contemporary Creative Economy and its Critical Discourse

The creative economy as a concept has become associated with the broader knowledge economy as a tool to understand urban regeneration and structural change in post-industrial urban areas across Europe (Banks, 2009; O'Connor, 1998).

Prior to the 1990s, however, the 'cultural industries' was the preferred term, with cultural policy and sociologists informed by the politics of contemporary cultural studies (Flew, 2012; Williams, 1958; Hall and Whannel, 1964). 'The Cultural Industry' was originally associated with the Frankfurt School of Critical Theory, and particularly the work of Theodor Adorno and Max Horkheimer (1977[1944]), who coined the phrase to pessimistically draw attention to the commodification of culture observed from the 1940s onwards in the United States. In turn, Bernard Miège (1989) and sociologists working in the area challenged the critique of mass culture centred upon new industrial technologies and the production of cultural symbols. Instead, in the context of post-1968 social movements, Miège argued for recognition of differences across the many sub-sectors of the cultural *industries*. Miège helped to popularise the plural term 'cultural industries' (as opposed to the singular 'The Culture Industry') to signal 'complexity, contestation and ambivalence' in the study of culture, maintaining that there were roles for culture beyond capital (Hesmondhalgh, 2012: 45).

The cultural industries were thus the precursor to the creative industries, however, the political agendas underpinning them initially diverged. Local socialism fed into early formations of the creative industries in the UK (Flew, 2013). In documents of Labour-led local councils from the 1980s, including Liverpool, Manchester and Glasgow, the cultural industries were mobilised to fill the economic hole left by mass de-industrialisation and manufacturing (ibid.: 15). Bianchini and Landry (1995) were the first to articulate a coherent formulation of the 'creative city', explicitly aligning the cultural industries with urban regeneration. Creativity subsequently gained traction as a doctrine for left-leaning and right-leaning policy-makers world-wide, with US academic Richard Florida (2002) leading a fervent boosterist approach. Fusing the creative economy with potential for inward investment, the creative class hypothesis spatially nests cities and creative talent with concentrated and enhanced economic prosperity (ibid.). 'Creative Class' focus rests on attracting the top 30 per cent of earners. While creativity still features in public policy and urban development discourses, in recent years language has shifted to a celebration of innovation, albeit with similar kinds of issues applying (Hesmondalgh & Baker, 2011: 4).

But it was the UK New Labour governments of Tony Blair and Gordon Brown that first popularised the role of the creative industries within policy discourse. After 1997, developing policies to promote the arts, media, design and software sectors (Flew, 2012: 3), New Labour drove a new creative economy agenda with output mapping exercises in 1998 and 2001. These mapping exercises were criticised for failing to address regional and local variation, or failing to compare different nation states (Pratt, 2004; Flew, 2013: 21). Some of these inadequacies were addressed in the Second Creative Industries report, the release of which coincided with the establishment of the Scottish Assembly, Welsh Parliament and Regional Development Agencies (RDAs) in 1999, which created new forms of regional governance. In a crucial move, traditional cultural industries were wrapped together with rapidly expanding sub-sectors, meaning that the creative sector could be presented as an area of special growth. The creative industries were positioned so they connected the non-monetary value of culture – such as personal fulfilment and self-actualisation – with measurable business outcomes. That is, combining profits with purpose. After the UK championed the creative industries, there followed a subsequent uptake by international bodies, including the United Nations Educational, Scientific and Cultural Organization (UNESCO), the United Nations Conference on Trade and Development (UNCTAD) and the World Trade Organisation (WTO), along with nation states both developed and developing (Flew, 2012: 2). A report by *Digital Britain* in 2009 (DCMS and Department for Business, Innovation and Skills) positioned Britain as a global centre for creative industries in a digital age. Yet indeterminacy of how to define the breadth and depth of the industries and occupations in the creative economy remains an issue worldwide (Hesmondhalgh, 2012: 217; Prince, 2014; Pratt, 2005).

As Hesmondhalgh and Baker (2011: 3) argue, the adoption of the term 'creative' was aided by its largely benign associations given the prestige attached to art and knowledge in almost all societies and civilisations. Social scientists popularised the notion of creativity from the 1950s, however human relations management and economists, specifically theories of endogenous growth, were highly influential in stimulating government planning in the creative economy (ibid., also see, for example, Guilford, 1950; Koestler, 1964; Amabile, 1996; Peters, 1997; Aghion, et al., 1998). Ideas on creative policy orientated towards technological advancement, and therefore innovation, were directly influenced by business agendas: 'In an era where government increasingly took many of its cues from business fashion, think tanks were soon portraying creativity as a key source of prosperity in the post-industrial city' (Hesmondhalgh and Baker, 2011: 4; also see Landry and Bianchini, 1995).

In recent years mounting criticism has been levelled at creative economy policy agendas. The creative economy has been argued to be driven by a skills and employability agenda, which pays little attention to the negative aspects of creative labour markets (Banks and Hesmondalgh, 2009). Rather than anti-conformist or subversive, artistic work can in fact be characterised by conditions compliant with modern capitalism such as flexibility and tolerance of inequality (Menger, 2006: 6). For many, creative labour has become casualised, with individuals often working in isolation on a project-by-project basis, assuming the economic and social risks and costs of creative output (see Bain, 2005). The creative city script is widely viewed as neglecting the power relations, such as self-disciplining and risk-taking, that confront members of the so-called creative class, and also the underlying power relations that impact upon inequalities in creative labour market access and divisions of labour (Oakley, 2006: 265). The result is that the values of enterprise culture and individualism are seen to triumph over 'social justice and collective solidarity' (Flew, 2012: 106).

In particular, a growing body of research has emerged which is highly critical of the brand of creative cities discourse championed by Florida and Landry. For instance, Terry Flew critiques the account of urban city centres as unique incubators, arguing instead that the sector draws upon 'the whole range of resources in a city' including those from the suburbs and wider city-region (2012: 7). Meanwhile, recognising a broader spatial unevenness in creative cities ideas, Tim Edensor et al. (2010: 5) warn against the instrumental uses of arts and culture that 'venerate the metropolis' as the site of consumption and production to the exclusion of other spaces, including the rural. Issues of social justice have rightly been highlighted, in particular the production of a co-dependent service class, and gender and income inequalities (Peck, 2005; Leslie and Catungal, 2012). Challenging the 'creative class' approach, Ann Markusen (2006) has argued that artists, a 'core' sub-sector of the so-called creative class, are actually likely to be relatively less affluent. Moreover, rather than the creative class developing weaker community links, artists – so-called 'core creatives' – may play more active roles in their neighbourhoods (ibid., 1923, 1937). This reminds us to think carefully about the specificity of the sub-sector and context, intended purpose and unexpected outcomes of any creative work. In the next section to what extent community is a significant and useful concept in creative economy discourse to understand local, inter-local and place-based impacts is the subject of further exploration.

Community and the Creative Economy

As Raymond Williams (1976: 66) identified, no other term of social organisation carries such positive connotations of collective solidarity as 'community'. The concept of community has been the source of extensive academic and political attention. It can refer to a group of people with shared values, a shared sense of place, identity or politics, shared interests, and communities of practice. But with renewed academic, policy and practice-based interest in the potential of community, has come critique of how the concept is wielded. Community has favourably been associated with building social capital, examining social networks, civic participation, trust and cohesion. From Durkheim's 'anomie' to Putnam's notion of 'bowling alone', scholars have argued that the loss of community can result in a breakdown of social networks and identity, with resultant social impacts such as increased levels of anxiety and depression (Putnam, 2000; Durkheim, 1893). Participatory, action and activist-orientated research emerged from the 1970s –1990s, experimenting with new ways of representing the 'other' (Goodson and Phillimore, 2012; Rose, 2001). These research approaches shared an aim to make sense of increasing levels of difference experienced in many communities around ethnic and cultural diversity and wealth inequalities. Significantly, community has often been positioned in a state of tension with capital. Whereas in certain disciplines and schools of thinking this has been viewed in a positive light, in others the concept has been viewed with suspicion. In economics, for instance, community has adversely been identified with growth-limiting activities, such as blocking of change (Olson, 1965; Amin and Roberts, 2008). The discursive positioning of community is highlighted by Miranda Joseph (2002: 2) who dismisses the term as necessarily autonomous and antipathetic to capital, arguing 'community functions in complicity with "society" enabling capitalism and the liberal state'. Interconnected issues of empowerment, participation, decision-making and impact are part of this contested field of research and practice (Pain and Kindon, 2007; Cooke and Kothari, 2001). This edited collection through the various case studies investigates how academics, practitioners and policy-makers have approached the *value* of community and community research in relation to the creative economy.

i) Tacking Social Exclusion

In the UK, New Labour replaced the word poverty with the concept of 'social exclusion'. A significant intellectual shift was ushered in, by which the ability for people and areas to contribute economically became the key measure of their value (see Hall, 2013). DCMS concluded that arts, sport and cultural and recreational activity could be used to support regeneration at the neighbourhood scale. The belief was that creative and sports activities have positive impacts because they can:

- appeal directly to individuals' interests and develop their potential and self-confidence;
- relate to community identity and encourage collective effort;
- help build positive links with the wider community;
- be associated with rapidly growing industries (DCMS, 1999: 8).

Cultural and sporting activities were framed as improving health, crime, employment and education indicators in deprived communities, which, in turn, enabled citizens to be more economically productive (Jermyn, 2001) and therefore of greater value to society. This characterises New Labour's 'third way' assumption that economic growth will create social inclusion through 'trickle down' processes linking creative economy policies with local communities. Underpinned by an economic growth agenda, the DCMS (1999) report was a manifestation of New Labour's wider approach on cultural policy, which failed to measure the crucial non-monetary elements of cultural and creative activity, for instance volunteering practices (Gibson-Graham, 2006: 60).

Evans and Foord (2000) have shown that the origins of community arts practices during the 1970s in the UK and in Western Europe coincided with the first wave of major youth and structural unemployment and the subsequent rise of urban regeneration policy. Process-based methods in participatory creative and cultural activity here take on a functional purpose to draw people and places into an active civic sphere (Goldbard, 2006, 2009; Jermyn, 2001). In the UK creative activities are evaluated in accordance with social inclusion targets where sociality fostered in the spaces of art interaction is tacitly assessed as building social and cultural capital (see Hall, 2013). In this strand of the creative economy, cultural policy intersects with social policy. The value of the creative economy thus combines economic growth, urban regeneration, aesthetic quality, and social inclusion of marginalised and diverse communities at a neighbourhood level (see Oakley, 2006, 2009). Drawing connections between economy, innovation, revitalisation, sociality and citizenry, creative practice performs a political function by activating local community participation.

ii) Communities of Becoming and Belonging

Capturing the flux and multiplicities of community is a core strand of geographical engagements with cultural and creative activity that emphasises *becoming* and *belonging*. This body of work is characterised by a non-essentialising sense of community. For example, public art projects have been investigated as a vehicle through which 'sense of community' can be fostered (Hall and Robertson, 2001: 10). Sense of community in this context refers to 'an awareness of a social body occupying a shared space with connections stemming from some common identity, values or culture' (ibid.). By making forms of community in arts practice visible, the intent is to facilitate 'the emergence of social bodies', promoting community development and cohesion which echoes Putnam's work on social capital (ibid.).

Notions of belonging are also used to reflect the experiences of attachment, and desire for recognition. This is something Hall (2013) has examined in regard to people with learning difficulties in arts-based making and gifting. Instead of framing participation in cultural and creative activity as a conduit from social exclusion to social inclusion, Hall explores crossings between non-disabled and disability-exclusive spaces and the resulting potential for connection and belonging. Thus the concept of belonging offers an alternative from the fixed binary of exclusion/inclusion (see Antonsich, 2010). Probyn (1996: 19), for example, has revealed that *belonging* incorporates *becoming*; that is both being and yearning for 'some sort of attachment, be it to people, places or modes of being'. In doing so, a fixed or stable understanding of belonging to a community is unsettled.

Taking a non-essentialising understanding of community even further, Gillian Rose develops the notion of a 'community of lack' (1997). Drawing upon community arts practices, Rose attempts to challenge senses of identity that are predicated on stability and accord, and defined in relation to 'what is positioned as without' (ibid.: 2; Carter et al., 1993). Rose captures an important sense of porosity and flux in a spatial organisation of identity, which 'can acknowledge partial and changing membership; contingent insiderness; uncertainty, loss and absence' (1997: 14). Nevertheless important issues arise from labelling particular groups as lacking, disadvantaged or hard-to-reach. As Durose et al. (2013: 5–6) have observed, labelling groups relies 'on constructing a dichotomy between the mainstream and the marginalised'. Durose et al. give pause for reflection on the power-differentials of multiple stakeholders involved in local policy and practice-based interventions. We are further reminded of the agency of 'marginalised' individuals and groups, including those who may choose to opt out or remain on the peripheries of mainstream culture (ibid.).

iii) Agency, Networks and Inter-localism

The agency of communities in the creative economy is an emerging concern in recent work on urban and international development. More sustainable and inclusive community development practices are presented by engaging art and cultural activity to develop residents' skills, talents and needs in post-industrial cities (Margulies Breitbart, 2013: 62). The rise of ethnic neighbourhoods as sites for the commodification of ethno-cultural diversity and consumption is the subject of Aytar and Rath's (2012) volume which traces globalization 'from below'. With a shared focus on methods and practices for growing the creative economy, Barrowclough and Kozul-Wright (2008) open up the research lens to communities in developing countries. Yet, Ash Amin (2005) draws attention to the peculiar tension that at a time of greater connectivity and flows between places in different global networks there has been a return to policy directed towards restoration of local community and spatially enclosed places. The limitations of a spatially-bounded sense of community are signified by mobilities of people and products, particularly in the creative sector which often incorporates multi-stage and place networks on the value chain of cultural forms from production to consumption. Moreover, new forms of communication networks (e.g. Skype, social media, mobiles), along with cheaper air travel and new media channels means that that for many people experience of community is increasingly inter-local.

While community is not explicitly discussed (although cf. Warren and Jones, in press), a body of literature on cultural intermediation has analysed the connections between creative producers and the marketplace. Bourdieu (1979) used the term cultural intermediaries to identify the emergence and dynamics of a new kind of actor that arose in an era of expanding mass culture in 1960s France. This included TV producers, critics and journalists. Usage of the concept of cultural intermediaries has evolved to reflect the 'significant changes brought about by the growth of workers involved in the production and circulation of symbolic forms' (Negus, 2002: 501). Intermediaries 'identify, privilege and seek to stabilise' economic activities so they can be measured (O'Connor, 2015: 2; Jessop, 2005: 145) and are also observed to work as state, non-state and hybrid actors. As Taylor (2013: 2) has noted for intermediary agents operating between culture, policy and industry, the role reproduces 'prevailing economic relationships' and simultaneously drives the emergence of new associational relationships'.

Place Specificity, Policy and Practice

The practice of bringing together creative economy and communities is not an abstract one. Real people have to meet (whether this is through electronic media or face-to-face) to discuss, plan and enact these agendas. The locations where this interaction happens are crucial to the way that community, policy and practice are co-constituted, which is where the concept of *place* is a valuable conceptual frame. Geographers have gone to considerable effort to conceptualise the role that place plays within social processes (Tuan, 1977; Relph, 1976; Massey, 2004, 2005; Cresswell, 2008). Put at its simplest, place is a space with a meaning. Place is therefore a relational concept, constructed through the meanings that different individuals layer onto the world around them.

Place is not uncontroversial, however. The idea of place being somehow more 'authentic' than the anywhere spaces of globalised capitalism (Relph, 1976) raises questions about what an authentic place attachment might look like. Indeed, Harvey (1993) has suggested that the discourse of authentic places, with real meaning to the people who live there, runs the risk that the forces of capitalism will seek to use *place* as another tool in accumulation and exploitation, trading off nostalgia to sell a faux-authenticity. Does the continuing popularity of Budapest's 'ruin bars' speak of a locally meaningful legacy of post-communist collapse or is it simply that an apparently authentic legacy is a highly marketable commodity for attracting international tourists? Can both elements coexist?

This ties neatly into questions around specific place-marketing strategies pursued by cities and regions, seeking to emphasise the unique selling points of one location over another (Kearns and Philo, 1993). The idea of place as selling point has strong resonance within creative city discourses. The Florida (2002) prescription is in providing facilities that attract the creative classes to cities – a thriving arts and cultural scene being part of this. Even before Florida codified this idea, cities flocked to invest heavily in cultural regeneration schemes, with art galleries, concert halls, museums and so forth being used as anchors in major redevelopment projects: Bilbao's Guggenheim; Amsterdam's Muziekgebouw; the National Aquarium in Baltimore and so on. Of course, as Peck (2005) has noted, the danger with these attempts to use cultural facilities to enhance place identity is that they lapse into a new form of placeless generic. Analogues of Bilbao, for example, have sprung up in many places: does siting a new opera house (Oslo) or branch of the Victoria and Albert Museum (Dundee) on a waterfront really do much to enhance the authentic place qualities of these cities? The same could also be argued for more grassroots manifestations of 'authentic' local cultures – the aesthetic of mismatched tables and chairs, quirky crockery choices and fresh pastries served in independent cafes showing off exposed brickwork can be found across north America and western Europe. To what extent can these be said to be grounded in the specifics of a place?

Such criticisms set place up as being somewhat conservative (see Massey, 1994). Place can thus be read as a veil concealing manifestations of global culture, or as an ignorant, inward-looking resistance against progress (quite literally keeping citizens 'in their place'). Clearly there is some merit to this critique. Nonetheless, despite the best efforts of globalising capitalism, places *are* unique and different – no one is going to mistake Amsterdam for Aberdeen, Guangzhou for Gdansk – and the intersection of community and creative economy happens *differently* in those places. One can read this through de Certeau's (1984) notion of tactics and strategies. While the strategies of global capitalism drive toward a singular culture of consumption, the characteristics of different places (driven by their histories, cultures and inhabitants) act as a tactic of resistance against the production of a global generic. A concrete, if not particularly positive, example of this can be seen in the closure of The Public, an arts centre in the English midlands town of West Bromwich. The gallery cost over £50m to build and yet never seemed to connect sufficiently with local residents to attract a meaningful audience. Meanwhile, the geographic location of West Bromwich and the town's lack of other facilities that might attract tourists meant The Public never became a destination for people from outside the region (Jones and Evans, 2013). Faced with negligible audiences and a cost to the local authority of an estimated £30,000 per week to keep it open (BBC, 2013), the gallery was closed in late 2013, with relatively little fanfare. West Bromwich, it turns out, was not a place where the particular kind of culture produced by The Public could thrive.

Thus place is more than a quality of city marketing. But it is also more than simply a backdrop against which tensions between creative economy and community policies are played out; cultural policy, community and place are co-constituted. Perhaps one of the most significant manifestations of this has come in the way that cultural policy has become deeply entangled in notions of social exclusion and the spatial concentration of poverty. Simply living in the same geographic location does not make people members of the same community, nonetheless, when it comes to discussions of poverty and social exclusion, there tends to be an implicit assumption of geographic proximity creating community. People tend to socialise within income and cultural groupings and these groupings frequently cluster geographically. Of course, this is not simply a question of poverty: Kensington and Chelsea has become home to the well-heeled global elite, just as Cheshire's Alderley Edge has become a magnet for extraordinarily wealthy Premiership footballers. One can similarly identify areas with strong clustering by cultural heritage: Pakistani Muslim communities in parts of Birmingham, for example. Meanwhile, statistical measures such as the UK's index of multiple deprivation (Noble et al., 2006) can be used as tools to identify *areas* (as distinct from *places*) in need of intervention. Thus in evaluating the demographic reach of, for instance, a publicly subsidised exhibition, visitor data collected using postcodes can measure the extent to which the audience is coming from neighbourhoods considered deprived. This, in turn, becomes a proxy for participation by 'socially excluded' communities.

Notions of participation are important here because they tend to lead us back toward considering communities in terms of the deficit model. Failed places need to replace failed communities with good ones. If you participate you will become part of a good community. Therefore if policy fails, it is implicitly a failure of community rather than one of policy (Amin, 2005). This, of course, says little about the multiple challenges facing the specific places in which different communities live, whether this be lack of jobs, poor quality educational and recreational opportunities, environmental issues, and so on. As Hall and Robertson (2001: 19) note, the claims for public art having the capacity to translate abstract spaces into meaningful places are therefore somewhat problematic. A simple 'add culture and stir' model pays little attention to specific challenges facing specific communities in specific places.

The Structure of this Book

The chapters that follow seek to respond to this challenge of considering the interdependence and co-construction of creative economy, community and place. The book is divided into two parts. The first 'Creative practice, creating communities' focuses on the ways in which creative activities become a vehicle for fostering the development of particular kinds of communities. This part is opened by Shari Daya who explores how the act of artistic production can play a critical role in how individuals navigate the tension between economic value and performing community identity. While many scholars have emphasised the importance of materiality in an analysis of the economic, Daya's chapter responds to a challenge posed by Cook and Tolia-Kelly (2010) who note an overemphasis of materialities research focusing on consumption rather than production. Exploring beadwork producers in South Africa, the chapter moves beyond the conventional binaries that might depict beadworkers as either victims of global capitalism or, conversely, celebrate them as autonomous creative subjects. The emphasis on production brings the individual to the fore, relating stories of people walking the line between identities as creative actors and how their individual agency is bounded by their economic position. Beadwork is so ubiquitous in South Africa as to be almost invisible, but it demonstrates how ordinary craft cultures at the margins of formalised creative activities produce both economic value and communities. Beadwork thus serves a number of purposes for its producers; there is the economic value of artwork produced, but it also serves a critical role in identity formation. A key element is the ways in which place and community intersect with this productive activity. It is particularly telling in the case of migrant workers, coming from across southern Africa. Here beadwork connects producers to memories of place and communities, as they try to make ends meet while living a long way from home.

Ecosystem services is a broad concept which explores the value of ecosystems for continued human existence. It comprises several different components, from extractive value (for instance harvesting timber and minerals) through to cultural value (the aesthetic or spiritual qualities for humans of engaging with landscape). Acott and Urquhart's chapter explores these cultural ecosystem services (CES) as a way of thinking through how communities and cultures are shaped through interactions with the natural world. As part of a large research project into inshore fishing in north-west Europe, they have undertaken an exciting and innovative research approach using various different forms of photography. Photographs have become a key media through which we make sense of ourselves, our communities and our relations with the wider environment. Inshore fishing is a highly pertinent case study of connections between communities and landscapes as it has a very long history, yet is under tremendous pressure through environmental and legislative changes which are in turn reshaping local cultures. A combination of different photography-based research approaches are described, including a photographic audit of 75 coastal settlements, photo-elicitation interviews with local residents and exhibitions in six communities, showcasing commissioned work from a professional photographer. There is a risk with using professional images that such representations might tend toward aesthetically pleasing photographs of sunlit boats, rather than highlighting the life-threatening aspects of working with the sea. Acknowledging this risk, the chapter nonetheless highlights the intention to use the photographs not merely to document but to actively reshape community and policy narratives around cultural ecosystem services and inshore fishing.

Acott and Urquhart also highlight the fact that while terms like 'ecosystem services' can be alienating to participants, key ideas of nature and sense of place are highly resonant. Thus again, the idea of places having particular characteristics that might be preserved, cultivated or modified through creative intervention is at the heart of their discussion. Similarly, Dave O'Brien's chapter also reflects on the role of photography in shaping sense of place. Where Acott and Urquhart used a professional photographer to shape the imagery and resultant narrative, however, the *Some Cities* project described by O'Brien began from the point of professional photographers offering training to keen amateurs to create their own narratives. The broad theme of *Some Cities* was in celebrating the city of Birmingham, its diverse landscapes and communities. O'Brien was directly involved with the project in helping set up a participatory evaluation of the work. Evaluation is at the heart of how publicly funded cultural activities are regulated and yet, for all the discourse of placing the participation of communities at the heart of these activities, evaluation – the yardstick against which success is judged – tends to be top-down and externally imposed. Instead of documenting the ways in which participants engaged with the project, a participatory evaluation asks those participants instead to define *their own* terms of engagement with a project and to decide whether it succeeds against *their own* measures. This means that some of the benefits of engaging with a project that neither funders nor participants had expected – whether this be enhancing civic pride, new social networks, confidence to engage with community and place – can become key measures of a project's success.

Through a crowdsourcing website the *Some Cities* project has used the democratising potential of new photographic technology to create a powerful narrative of connection between the people of Birmingham and their city. It plays with the ways in which a city-wide place-based identity can be crystallised through a community of amateur creative producers. O'Brien's cautiously positive review of the role a well-constructed creative project can play in connecting community and place, is echoed in the chapter by Antonia Layard and Jane Milling. Layard and Milling use a legal perspective to investigate the ways in which creative place-making can be undertaken. Taking a deliberately narrow definition, place-making is investigated in terms of how public spaces are planned, designed and managed. Given that the legal framework is the same across England, they raise the question of why some groups feel more able to engage in place-making projects than others elsewhere in the country. Theories of legal consciousness explore how shared meanings are stabilised between groups and how these are then institutionalised through the creation of law. Drawing on Levine and Mellma (2001) they acknowledge that while the law is always present, its importance to individuals can vary over time.

The idea of acting 'under the law' is key to how Layard and Milling interpret processes of creative place-making. Rather than breaking the law, people working 'under the law' reposition themselves, conforming to a sense of a higher set of laws, attempting to create a different social and moral imaginary. In essence these 'under the law' practices attempt to shift how legal frameworks deal with particular issues. Exploring three case studies, they investigate different tactics by community groups trying to change elements of place. Newcastle Elders Forum, for example, undertook detailed surveys as well as artistic interventions (poems, skits) attempting to lobby for changes to the urban fabric to make it friendlier for older people. Northern Youth, conversely, became so frustrated at the lack of attention paid to their agenda promoting the interests of 8–25 year olds that as part of a delegation to the European Citizens' Panel they undertook a street protest in Brussels. The People's Republic of Stokes Croft meanwhile have seen their intention to set up a creative and cultural quarter in a struggling part of the city, at first opposed by the city council (including via a court case), but later influencing local policy about the neighbourhood.

Layard and Milling argue that acting 'under the law' – actively attempting to reposition formal legal structures to suit a particular agenda – opens the door to more creative place-making by creative communities. Of course, they note that one group's vision of how a place should be does not necessarily suit everyone living there and that there is a danger that different visions of place can be squeezed out – not everyone in Stokes Croft, for example, is enamoured of the graffiti culture that is such a visible influence on the street scene. Thus we cannot uncritically celebrate such attempts to influence policy.

As we move into the second part of the book, 'Policy connections, creative practice' the importance of policy and the role that formal policies play in shaping culture, community and place are brought to the fore. This part opens with a chapter by Ginnie Wollaston and Roxanna Collins. Both are professional policymakers working in the Culture and Visitor Economy Service at Birmingham City Council. What they offer is an insight into the interlocking set of initiatives that shape cultural policy within the city. The particular focus is on a pilot scheme which created a pot of cash for cultural commissioning in three relatively deprived neighbourhoods suffering from a variety of socio-economic problems: Balsall Heath; Castle Vale and Shard End. In part the pilot scheme was an attempt to find ways of engaging peripheral communities who are not well served by the large cultural institutions located within the city centre. The chapter introduces the idea of the *C2* approach developed by Gillespie and Hughes (2011) to engage the creative energies of communities into creating positive visions for change and finding ways to realise these. Although these were not strictly speaking participatory budgeting exercises (an external panel judged bids for resource), during the year-long pilot a number of locally-led projects were initiated. One major conclusion, however, is to raise a concern about the capacity that is available at local level for undertaking more active engagements in the process of commissioning and delivering cultural and creative activities at the neighbourhood scale. Dependence on a strong infrastructure of local creative intermediaries to facilitate these activities is particularly significant at a time when funding for these actors is being reduced.

Wollaston and Collins' chapter forms an interesting contrast to that by O'Brien in the ways in which evaluation is considered. In Birmingham's cultural pilot a significant proportion of the budget was set aside for undertaking both a qualitative survey of participants and organisers as well as more conventional quantitative studies of audience numbers. This partly reflects the kinds of pressures put onto local authorities to be able to demonstrate the value of their work, with a rather narrow definition of value. Changing this culture of course means changing some of the legal and governance frameworks within which local policymakers operate – though whether council officials would be comfortable working in the kind of 'under the law' mode that Layard and Milling discuss is an intriguing question!

Much of the discussion around culture and policy is implicitly urban in focus. Urban and rural communities often face quite different challenges, particularly in terms of communication and connectivity. These issues are explored in Liz Roberts and Leanne Townsend's chapter on rural creative economies. It is easy living in a city to take the rapid communication of the digital age for granted, but infrastructure such as fibre optic broadband and fast mobile data connectivity are unevenly distributed, focusing on the more profitable areas of population concentration in the cities. Therefore in policy terms, Roberts and Townsend reflect that the rural creative economy is thought of in terms of 'potential' rather than actually delivering the promised gains for communities, despite bringing an estimated £500m into the UK rural economy each year. In rural areas, the creative economy is disproportionately represented by self-employed and microbusinesses, acting very much in contrast to the Florida (2002) inspired discourse of clustering and buzz being essential to this kind of activity.

From a policy point of view there have been initiatives such as the Rural Community Broadband Fund, that attempt to bring the benefits of greater connectivity to rural creative entrepreneurs. Nonetheless, there is an acknowledgement among policymakers that universal provision, especially for remote areas, simply will not happen. The creative practitioners interviewed by Roberts and Townsend had a variety of approaches to tackling their isolation from these communication networks. There was a common theme of frustration that simple tasks such as exchanging digital files can take hours rather than seconds of effort. Similarly the importance of face-to-face meetings in terms of securing contracts for work often meant many hours of travelling. Nonetheless, the inspirational qualities of rural landscapes represented a major pull to these practitioners, despite the practical difficulties caused by their location and the inadequacy of policy frameworks to provide the infrastructure that city dwellers take for granted. Indeed there is a clear failure of cultural industries policy, with its focus on high tech, to take into account the low-tech, low-speed realities of the everyday rural. Thus we see a very stark illustration of the ways in which the place where creative economy is *done* has a profound influence on connections within professional communities and the way creative businesses can operate.

A centre/periphery divide can also be seen in Monika Murzyn-Kupisz and Jarosław Działek's discussion of museum and library provision in post-Socialist Poland. The post-Socialist context sets up a very particular set of challenges as older organisational structures start to be challenged by the demands of a conventional capitalist economy and the purpose of established cultural institutions starts to change. This is being seen quite acutely in threats to the small branch museums and libraries existing in smaller towns at some distance from the main cities as funding structures are re-arranged and the mission of these institutions altered. The significance is in an explicit policy focus on the role that these institutions can play in enhancing social capital of communities, particularly in creating spaces of encounter and dialogue within and between communities. These insights are especially important in a post-Socialist context where the web of communal and voluntary associations are much weaker in Poland than elsewhere in Europe, creating challenges to strengthening community links beyond the immediate family group. The problem is in delivering on this new mission within an existing set of physical infrastructures – small branch libraries, for example, rarely have the space for meeting rooms and hosting minor exhibitions. Limited availability of funds has meant infrastructure upgrades have focused on core functions; investment designed to enhance capacity to bring communities together has tended to be restricted to the major institutions in cities such as Krakow, leaving provincial areas underserved.

Much of the innovative work that has happened with a new generation of library and museum managers in Poland has been delivered through accessing European Union co-finance. The final substantive chapter in this collection is by Paul Long and Steve Harding who investigate a Europe-wide project looking at the policy framing that connects people and places into the creative economy. This is an important topic to EU policymakers as the creative economy is seen as a key engine of growth across the continent, with a variety of policy instruments to this end outlined in a Green Paper *Unlocking the Potential of Cultural and Creative Industries* (European Commission, 2010). The *Cross Innovation* project, of which Long and Harding are a part, seeks to promote policies, research and practice that foster exchange between the creative industries and other sectors. In part, this activity is a reaction to the perceived failure of the Lisbon Agenda of 2000, which aimed to make Europe a leader in the knowledge economy in order to drive sustainable growth and social cohesion.

Long and Harding discuss *Cross Innovation* through the lens of cultural intermediation, where actors seek to join up creative and cultural activity with wider communities. This process of connection is key to cross-boundary projects, such as the serious computer game that forms part of cargo managers' training at Schiphol Airport, or *Kolonien* in Stockholm, a collaborative shared working space facilitating companies' concept development, design and marketing. Intermediaries are at the heart of developing these connections, but face multiple challenges, not least that in many cases cooperation within the sector does not necessarily start with commercial profit and yet there is policy pressure to ensure creative enterprises can drive economic growth. There is a clear emphasis, however, that there cannot be a one-size solution to these issues, with place-specific challenges facing the wider EU project. Nonetheless, the ideas being championed in *Cross Innovation* act as a fascinating and sometimes contradictory site for remaking identity and citizenship across the EU through the power of cultural and creative policy.

In the concluding chapter we identify the cross cutting themes that emerge from this book, in particular looking at the role that the idea of 'place' plays in connecting together community, policy and creativity. The key here is not to think of place as a singular, unified scale, somehow more 'authentic' because it is closer to 'communities'. Instead place is relational, cutting across scales and challenging the one-policy-prescription-fits-all model of community + culture = economic success. In drawing together the lessons from across the chapters in the book we reflect on a tension between an economic geography emphasis on clustering driving success and the ways in which creativity and communities often operate at the margins, in peripheral places both within and without cities. At its heart this book reflects on the challenges being posed by an implicit disconnect between the ways in which policy is formulated, creative practice enacted and communities brought into being. Places are not simply where these dis/connections happen, but instead become an active agent that needs more explicit consideration when investigating ways to improve those connections.

References

Adorno, T. and Horkheimer, M. (1977[1944]) *Dialectic of Enlightenment.* Continuum: New York.

Aghion, P., Howitt, P., Brant-Collett, M. and Garcia-Peñalosa, C. (1998) *Endogenous Growth Theory*. MIT Press: Cambridge, MA.

Allmendinger, P. and Haughton, G. (2012) Post-political spatial planning in England: a crisis of consensus? *Transactions of the Institute of British Geographers* 37(1), 89–103.

Amabile, T. (1996) *Creativity in Context.* Westview Press: Boulder, CO.

Amin, A. (2005) Local community on trial. *Economy and Society* 34(4), 612–633.

Amin, A. and Roberts, J. (2008) *Community, Economic Creativity, and Organization*. Oxford University Press: Oxford.

Antonsich, M. (2010) Searching for belonging: an analytical framework. *Geography Compass* 4(6), 644–659.

Aytar, V. and Rath, J. (2012) *Selling Ethnic Neighbourhoods: The Rise of Neighbourhoods as Places of Leisure and Consumption*. Routledge: London.

Bain, A. (2005) Constructing an artistic identity. *Work, Employment, and Society* 19(1), 25–46.

Bakhshi, H., Freeman, A. and Higgs, P. (2013) *A Dynamic Mapping of the UK's Creative Industries*. NESTA: London.

Banks, M. (2009) Fit and working again: the instrumental leisure of the Creative Class. *Environment and Planning A* 41(3), 668–681.

Banks, M. and Hesmondhalgh, D. (2009) Looking for work in creative industries policy. *International Journal of Cultural Policy* 15(4), 415–430.

Barrowclough, D. and Kozul-Wright, Z. (2008) *Creative Industries and Developing Countries: Voice, Choice and Economic Growth*. Routledge: London.

BBC 2013 *The Last Day of the Public Art Gallery in West Bromwich* http://www.bbc.co.uk/news/uk-england-birmingham-24973336, accessed 29 January 2015.

Bourdieu, P. (1979; trans. 1984) *Distinction*. Routledge: Oxford.

Carter, E. Donald, J. and Squires, J. (1993) *Space and Place: Theories of Identity and Location*. Lawrence & Wishart: London.

Cook, I. and Tolia-Kelly, D. (2010) Material geographies, in Hicks, D. and Beaudry, M. (eds) *Oxford Handbook of Material Culture*. Oxford University Press: Oxford, 99–122.

Cooke, B. and Kothari, U. eds (2001) *Participation: The New Tyranny?* Zed Books: New York.

Cresswell, T. (2008) Place: encountering geography as philosophy. *Geography* 93(3), 132–139.

DCMS (1999) *Policy Action Team 10: A Report to the Social Exclusion Unit*. DCMS: London.

DCMS (2006) *Developing Entrepreneurship for the Creative Industries*. DCMS: London.

de Certeau, M. (1984) *The Practice of Everyday Life*. University of California Press: Berkley.

Donald, B. Gertler, M. and Tyler, P. (2013) Creatives after the crash. *Cambridge Journal of Regions, Economy and Society* 6(1), 3–21.

Durkheim, E. (1893) *The Division of Labour in Society*. Free Press: New York.

Durose, C., Beebeejaun, Y., Rees, J., Richardson, J. and Richardson, L. (2013) *Towards Co-production in Research with Communities*. Connected Communities Discussion Paper, AHRC, Swindon.

Edensor, T., Millington, S. and Rantisi, N. (2010) Introduction: rethinking creativity: critiquing the creative class thesis, in Edensor, T., Millington, S. and

Rantisi, N. (eds) *Spaces of Vernacular Creativity: Rethinking the Cultural Economy*. Routledge: London, 1–16.

European Commission (2010) *Green Paper on the Potential of Cultural and Creative Industries* http://europa.eu/legislation_summaries/culture/cu0006_en.htm, accessed 29 January 2015.

Evans, G. and Foord, J. (2000) Landscapes of cultural production and regeneration, in Benson, J. and Rose, M. (eds) *Urban Lifestyles: Spaces, Places, People*. Balkema: Rotterdam, 249–56.

Flew, T. (2012) *The Creative Industries: Culture and Policy*. Sage: London.

Flew, T. (2013) *Creative Industries and Urban Development: Creative Cities in the 21st Century*. Routledge: London.

Florida, R. (2002) *The Rise of the Creative Class and How it's Transforming Work, Leisure, Community and Everyday Life*. Basic Books: New York.

Gibson-Graham, J.K. (2006) *Post-capitalist Politics*. University of Minnesota Press: Minneapolis.

Gillespie, J. and Hughes, S. (2011) *Positively Local: C2 a Model for Community Change*. Centre for Welfare Reform Policy Paper, University of Birmingham: Birmingham.

Goldbard, A. (2006) *New Creative Community: The Art of Cultural Development*. New Village Press: Oakland, CA.

Goldbard, A. (2009) Arguments for cultural democracy and community cultural development. *GIA Reader* 20(1), no pagination.

Goodson, L. and Phillimore, J. (2012) *Community Research for Participation: From Theory to Method*. Policy Press: Bristol.

Guilford, J.P. (1950) Creativity. *American Psychologist* 5, 444–454.

Hall, E. (2013) Making and gifting belonging: creative arts and people with learning disabilities. *Environment and Planning A* 45(2), 244–262.

Hall, S. and Whannel, P. (1967) *The Popular Arts*. Beacon Press: Boston.

Hall, T. and Robertson, I. (2001) Public art and urban regeneration: advocacy, claims and critical debates. *Landscape Research* 26(1), 5–26.

Harvey, D. (1993) From space to place and back again: reflections on the condition of postmodernity, in Bird, J., Curtis, B., Putnam, T., Robertson, G. and Tickner, L. (eds) *Mapping the Futures: Local Cultures, Global Change*. Routledge: London, 3–29.

Hesmondhalgh, D. (2012) *The Cultural Industries*. Sage: London.

Hesmondalgh, D. and Baker, S. (2011) *Creative Labour: Media Work in Three Cultural Industries*. Routledge: London.

Hracs, B. (2013) Cultural intermediaries in the digital age: the case of independent musicians and managers in Toronto. *Regional Studies* 49(3), 461–475.

Hracs, B., Jakob, D. and Hauge, A. (2013) Standing out in the crowd: the rise of exclusivity-based strategies to compete in the contemporary marketplace for music and fashion. *Environment and Planning A* 45(5), 1144–1161.

Imrie, R. and Raco, M. eds (2003) *Urban Renaissance? New Labour, Community and Urban Policy*. Policy Press: Bristol.

Jermyn, H. (2001) *The Arts and Social Exclusion: A Review Prepared for the Arts Council of England*. Arts Council of England: London.

Jessop, B. (2005) Cultural political economy, the knowledge-based economy, and the state, in Slater, D. and Barry, A. (eds) *The Technological Economy*. Routledge: London, 142–164.

Jones, P. and Evans, J. (2013) *Urban Regeneration in the UK, second edition*. Sage: London.

Joseph, M. (2002) *Against the Romance of Community*. University of Minnesota Press: Minneapolis.

Kearnes, G. and Philo, C. (1993) *Selling Places: The City as Cultural Capital, Past and Present*. Pergamon Press: Oxford.

Koestler, A. (1964) *The Art of Creation*. Macmillan: Oxford.

Landry, C. and Bianchini, F. (1995) *The Creative City*. Demos: London.

Leslie, D. and Catungal, J.P. (2012) Social justice and the creative city: class, gender and racial inequalities. *Geography Compass* 6(3), 111–122.

Levine, K. and Mellema, V. (2001) Strategizing the street: how law matters in the lives of women in the street-level drug economy. *Law & Social Inquiry* 26(1), 169–207.

Margulies Breitbart, M. (2013) *Creative Economies in Post-industrial Cities: Manufacturing a (Different) Scene*. Ashgate: Farnham.

Markusen, A. (2006) Urban development and the politics of a creative class: evidence from a study of artists. *Environment and Planning A* 38(10), 1921–1940.

Massey, D. (1994) *Space, Place and Gender*. Polity Press: Cambridge.

Massey, D. (2004) The responsibilities of place. *Local Economy* 19(2), 97–101.

Massey, D. (2005) *For Space*. Sage: London.

McRobbie, A. (2002) Clubs to companies: notes on the decline of political culture in speeded up creative worlds. *Cultural Studies* 16(4), 516–531.

Menger, P.M. (2006) Artistic labour markets: contingent work, excess supply and occupational risk management, in Ginsberg, V.A. and Throsby, D. (eds) *Handbook of the Economics of Art and Culture*. Elsevier: Amsterdam.

Miège, B. (1989) *The Capitalisation of Cultural Production*. International General: New York.

Negus, K. (2002) The work of cultural intermediaries and the enduring distance between production and consumption. *Cultural Studies* 16(4), 501–515.

Noble, M., Wright, G., Smith, G. and Dibben, C. (2006) Measuring multiple deprivation at the small-area level. *Environment and Planning A* 38(1), 169–185.

O'Connor, J. (1998) Popular culture, cultural intermediaries and urban regeneration, in Hall, P. and Hubbard, T. (eds) *The Entrepreneurial City: Geographies of Politics, Regime and Representation*. Wiley: Chichester, 225–239.

O'Connor, J. (2015) Intermediaries and imaginaries in the cultural and creative industries. *Regional Studies* 49(3), 374–387.

Oakley, K. (2006) Include us out: economic development and social policy in the creative industries. *Cultural Trends* 15(4), 255–273.

Oakley, K. (2009) The disappearing arts: creativity and innovation after the creative industries. *The International Journal of Cultural Policy* 15(4), 403–413.
Olson, M. (1965) *The Logic of Collective Action: Public Goods and the Theory of Groups*. Harvard University Press: Cambridge, MA.
Pain, R. and Kindon, S. (2007) Participatory geographies. *Environment and Planning A* 39(12), 2807–2812.
Peck, J. (2005) Struggling with the creative class. *International Journal of Urban and Regional Research* 29(4), 740–770.
Peters, T. (1997) *The Circle of Innovation*. Vintage: New York.
Pope, R. (2005) *Creativity: Theory, History, Practice*. Psychology Press: London.
Pratt, A. (2004) The cultural economy: a call for spatialized 'production of culture' perspectives. *International Journal of Cultural Studies* 7(1), 117–128.
Pratt, A. (2005) Cultural industries and public policy: an oxymoron? *International Journal of Cultural Policy* 11(1), 31–44.
Prince, R. (2014) Consultants and the global assemblage of culture and creativity. *Transactions of the Institute of British Geographers* 39(1), 90–101.
Probyn, E. (1996) *Outside Belongings*. Routledge: New York.
Putnam, R. (2000) *Bowling Alone: The Collapse and Revival of American Community*. Simon & Schuster: New York.
Relph, E. (1976) *Place and Placelessness*. Pion: London.
Rose, G. (1997) Spatialities of 'community', power and change: the imagined geographies of community arts projects. *Cultural Studies* 11(1), 1–16.
Rose, G. (2001) *Visual Methodologies*. Sage: London.
Taylor, C. (2013) Between culture, policy and industry: modalities of intermediation in the creative economy. *Regional Studies* 49(3), 362–373.
Tuan, Y.-F. (1977) *Space and Place: The Perspective of Experience*. University of Minnesota Press: Minneapolis.
Warren, S. and Jones, P. (in press) Local governance, disadvantaged communities and cultural intermediation in the creative urban economy', *Environment and Planning C* DOI 10.1068/c13305.
Williams, R. (1958) Culture is ordinary, in Mackenzie, N. (ed.) *Conviction*. MacGibbon & Kee: London.
Williams, R. (1976) *Marxism and Literature*. Oxford University Press: Oxford.

PART I
Creative Practice, Creating Communities

Chapter 2

Producing People: The Socio-materialities of African Beadwork

Shari Daya

Introduction

In the new urban park in Green Point, one of the positive local legacies of South Africa's hosting of the 2010 FIFA World Cup, a glittering snake is coiled in the indigenous groundcover, two iridescent sunbirds flash in the sun, and a giant ladybird rises in front of blocks of flats overlooking Table Bay. A new series of South African postal stamps features two very similar ladybirds, a Ndebele angel, a cell phone, several birds and a boxer with bright red gloves. At the busy intersection of Rhodes Avenue and the M3 highway, traversed by most journeys to and from the famous Kirstenbosch National Botanical Garden, young Zimbabwean men hawk indigenous flowers – bright bunches of strelitzia, agapanthus, and arum lilies. On a street corner in Kalk Bay, a seaside suburb 35 kilometres south of the city, a woman sits outside a popular bakery selling dolphins, sharks and whales along with red chilli peppers and hearts on key-rings. And at a craft-market in Rondebosch, a life-size lion is for sale. It has a light bulb inside to turn it into a glowing tawny brilliance at night.

All these objects are made from beads. With a few exceptions (the cell phone, boxer and angel, which incorporate textiles and other materials), they are made by threading tiny seed beads onto lengths of wire which are skilfully bent and twisted into the desired shape. These beaded objects are intricate, elegant and represent many hours' labour. A large figure can take weeks, even months, to complete. They are uniquely late-twentieth and early-twenty-first century artefacts in their combination of wire and glass beads. And they are everywhere. Beaded objects such as these are for sale at any site that tourists might be expected to visit in Cape Town, which means most of the southern suburbs and everywhere in the Central Business District (CBD) and its immediate surrounds. Most often, they are made and sold by informal traders, most of whom are migrants from other African countries. Beadwork is so ubiquitous that for locals, it goes for the most part almost unnoticed. The objects make distinctive gifts for foreigners, and can lend an African flavour to one's own home décor, but otherwise they form part of the background of the urban landscape. They are ever-present, cheap, pretty, and unremarkable.

Figure 2.1 Beaded ladybird in Green Point Urban Park
Source: S. Daya

To borrow an idea from Daniel Miller, informally traded beadwork in Cape Town has become 'blindingly obvious'. 'This implies that when something is sufficiently evident it can reach a point at which we are blinded to its presence, rather than reminded of its presence' (Miller, 2010: 51). For most people, most of the time, ubiquity makes things less rather than more interesting. Certainly, while Cape Town carries the title of 2014 World Design Capital, whose trademark yellow dots proliferate across the city, the producers of these ordinary beaded artefacts are seldom recognised as playing a role in the city's much-lauded creative economy. The starting point of this chapter is that the very ordinariness of beadwork (and more generally, the informal creative economy) in the city occludes the fact that it constitutes an important element in shaping people and social relations. Indeed the materialities of beadwork production and trade actually shape communities, both within local places and across global space.

Further, I argue that while the recent burgeoning of research into *things* in the social sciences has done much to enrich our understanding of consumption as a site of sociality, the realm of production, particularly in the global South, has been relatively neglected. Through the personal accounts of informal beadwork producers in Cape Town, I demonstrate the importance of taking production in Africa seriously as a site of social and cultural meaning. This neither denies the insights of consumption studies, nor does it call for a return to 'a simple productionism in which economic activity determines culture and the cultural context of commodities' (Mansfield, 2003: 180). Rather, it draws attention to the ways in which, when informal crafters in one African city talk about their work, they tell stories of new identities and new communities, as well as existing communities being sustained across time and space. This suggests that production, particularly creative production, should be recognised not only as a means of making things, but as a means of making people (Miller, 2010).

Consumers, Producers, and Geography's Materialist Turn

There has been in the past two decades or so, a great turning to *things* across the humanities and social sciences, leading several scholars to remark on a 'materialist turn', particularly within cultural geography. For some, such as Anderson and Harrison (2010), this shift begins to redress what they regard as an over-enthusiastic uptake in the 1980s and 1990s of social constructionism as the default interpretive paradigm, such that matter was 'no longer viewed as a problem relevant to humanistic criticism' (Tiffany, 2001: 75). Authority for materialist inquiry was instead ceded to science, with the cultural disciplines rendered dependent on the latter as the sole arbiter of the nature of things.

Re-claiming the space of materialist debate, social and cultural theorists are returning to questions of how and why matter comes to matter, and this has yielded a wealth of new stories and ideas about the nature of things. Alongside related work in anthropology and social theory more generally, geography has seen a reworking of its traditional materialist paradigms (Crang, 2005; Whatmore, 2006), along with many calls for – as well as critiques of – a 'rematerialisation' of the discipline (Jackson, 2000; Lees, 2002; Kearnes, 2003; Anderson and Wylie, 2009). One of the most prominent areas of research within this shift explores 'the social lives of things' (Appadurai, 1986). Concerned with commodities of all kinds, scholars have traced the journeys and social relations of fruit (Cook et al., 2004), flowers (Hughes, 2000), carpets (O'Neill, 1999), handicrafts (Green, 2008), mobile phones (Pfaff, 2010), shoes (Scheld, 2003), jeans (Crewe, 2008) and caterpillar fungus (Yeh and Lama, 2013), to name just a few of the 'things' that have been 'followed'. While geographers' theoretical approaches to understanding matter differ, sometimes significantly, there can therefore be no doubt that materialism has come to be widely embraced 'as a mode of practical and philosophical engagement' within the discipline (Kirsch, 2012: 434).

However, as several recent reviews have noted, this emerging body of work displays a strong tendency to privilege consumption over production as a site of sociality and cultural meaning. Cook and Tolia-Kelly (2010: 111), for example, remark on the relative absence not only of producers but also of 'designers, distributors, sellers, repairers, disposers, collectors, re-sellers, thieves, counterfeiters, etc.' in 'material cultural geographies'. Similarly, Mansfield (2003: 177) observes that 'recent literature on material culture of consumption, the culture industries, and business culture has not been able to address the dynamics of production', and Jackson (1999) notes that consumption tends to be privileged above production as the economic point at which 'culture' is created. While consumption has come to be read as a site of creativity, interpretation, and identity-formation, production, within the discipline of geography at least, is generally seen as primarily economic, less tangled in the intricacies of signification, sociality, emotion, and cultural meaning.

It is certainly true that scholars within and beyond geography have convincingly demonstrated the profound ways in which consumption shapes us as human beings. Studies of domestic spaces (Miller, 2001; Tolia-Kelly, 2004; Daniels, 2010), shopping (Miller, 2004; Colls, 2006), getting dressed (Woodward, 2007), eating (Jackson, 2010; Hayes-Conroy and Hayes-Conroy, 2008) and drinking (Miller, 1998) have opened up the multiple meanings and socialities generated by the purchase and use of all kinds of things. As Barnett et al. (2005: 4) point out, such research 'has demonstrated that everyday commodity consumption is a realm for the actualisation of capacities for autonomous action, reflexive monitoring of conduct, and the self-fashioning of relationships between selves and others'.

To call for more careful analysis of production practices is not to deny these contributions, nor is it to argue for a straightforward return to earlier, Marxist-inflected economic geographies that almost exclusively addressed issues of production and labour. Rather, the argument is that what is needed is a socially- and culturally-sensitive analysis of production that was lacking in earlier geographies of production and continues to be relatively neglected in contemporary consumption-focused studies.

This argument should not be read, however, as implying that producers are entirely absent from the new material cultural geographies. There are in fact numerous examples of producer-centred stories within this literature; indeed much of it is explicitly aimed at exposing conditions of production, usually in the global South, and the hidden connections in various commodity networks. Harvey is often credited with initiating this trend, via his call in 1990 for more politically responsible consumption. Although the field has been criticised as losing its radical edge (Leslie and Reimer, 1999; Hartwick, 2000; Castree, 2004), it is explicitly driven by a politics of connection and compassion, and Southern producers figure prominently as the objects of such empathy in many of these accounts.

As Benson and Fischer (2007: 802) argue, however, research into the experiences of these producers tends 'to either critique the economic exploitation and increased risk of global capitalism or celebrate [production practices] as local resistance to the world market'. That is, the framework for understanding producers in the South is predominantly one of economic development rather than one that deepens understanding of their ordinariness, their creativity, and the ways in which their production practices shape them as individuals, as well as their relationships with others in communities both proximate and stretched across distance. Often, the underlying narrative is one of deficiency – emphasising producers' vulnerability, poverty and oppression – or one of overcoming such deficiencies. Edjabe and Pieterse (2010: 5) make the point colourfully, arguing that Southern (specifically African) producers:

> [A]re either bravely en route to empowering themselves to attain sustainable livelihoods or the debased perpetrators of the most unimaginable acts of misanthropy. Explanations for these one-dimensional distortions vary from historical dependency perspectives, to the vagaries of the peddlers of neoliberal globalisation agendas, or to the glorious agency of dignified actors who persist with their backs straight, chin up despite the cruelties bestowed by governmental neglect and economic malice.

While this may represent something of an overstatement in the case of recent geographical accounts of Southern producers, there is certainly an emphasis in this literature on the ways in which famers (e.g. Hughes, 2000; Cook et al., 2004) and factory workers (Crewe, 2008) are exploited, pressured, and subjected to significant risk and uncertainty (both corporeal and financial) by more powerful economic actors. Similarly, other geographers have highlighted how even within 'ethical' trade networks, Southern producers such as agricultural and other small-scale producers can be 'commoditised', 'fetishized' (Goodman, 2004), 'patronised', 'alienated', and 'excluded' (Dolan, 2010) within global commodity networks.

Producers in these accounts are not constructed simplistically as the victims of globalisation or neoliberalism. Collectively, however, these studies are indicative of a powerful narrative of exploitation and victimhood that leaves little room for richer stories about producers as people in their own right, defined not by their relationship to the consumers of their goods but by their own sense of self, their communities of family, friends, and acquaintances, and their more-than-economic activities.

The parallel narrative of agency within this body of research, as both Benson and Fischer (2007), and Edjabe and Pieterse (2010) remark, goes some way towards tempering the dominance of stories of exploitation. This is evidenced by the emphasis on 'empowerment', 'struggle', and indeed 'agency' in many accounts of production in the global South. While this amounts to a more positive representation of producers, such accounts continue to imagine Southern producers in fairly narrow terms, theorising their sense of self primarily through the economic success or otherwise of their work, and providing little sense of them as rounded people.

Two recent accounts illustrate how a more socially and culturally sensitive account of Southern producers may be approached, avoiding the temptation to view their work solely through the lenses of exploitation or empowerment, and drawing instead on notions of desire and enchantment to re-imagine the practices of production. The first study, Benson and Fischer's (2007: 802) study of non-traditional agriculture, cited above, began with the intention of 'uncover[ing] exploitation at the hither side of the commodity chain' but, crucially, the authors found farmers' stories to be more ambiguous than this framework allowed for. Their analysis therefore explores producers' economic engagement instead through a framework of desire which, while 'not elid[ing] crucial questions of power [and, indeed, empowerment], agency, and inequality', captures more human dimensions of existence than dominant frameworks of need, livelihood, oppression and/or resistance. Incorporating 'cultural values, moral models, and hopes for the future', this study 'resist[s] the urge to reproduce the simple lopsided equation of southern need versus northern desire' (Benson and Fischer 2007: 814). The result is a portrayal of Southern producers as simultaneously more ordinary and more culturally rich than most studies recognise.

The second study, authored by Ramsay (2009) also trains a novel lens on Southern producers, here Swazi craft producers, in a theorisation of production, marketing, selling and purchasing as forms of enchantment. Noting that geographies of 'following' that aim to connect producers and consumers tend to overlook – in their focus on exploitation – 'those encounters which fall beyond the "commodity" status of an object' (Ramsay, 2009: 200), Ramsay elides questions of economic oppression and empowerment in favour of an understanding of producers as weaving a kind of magic through a range of strategies. These include 'inviting tourists to watch souvenir production processes' (Ramsay, 2009: 204), and offering 'entertainment and explanation' to captivate potential buyers. Rather than suffering, struggling or overcoming, Ramsay's producers appear as magicians, conjurers, alchemists and storytellers, infusing objects with memories and stories that entice and enchant.

These two accounts demonstrate that the practices of production (including such activities as designing and selling) are as intrinsically social and cultural as consumers and their purchasing, collecting, and gifting behaviours. The producers discussed by Benson and Fischer (2007) and Ramsay (2009) are interesting as people in their own right and in multiple dimensions, not merely as economic actors. In what follows, I turn to the informal creative economy of beadwork in Cape Town to further explore some of the ways in which production may be seen as generative of communities, identities and meaning, as we now readily accept to be true of consumption.

Beadwork Producers in Cape Town

The analysis that follows is based on around 80 interviews with informal beadwork producers and traders working in the central city and the suburbs of Cape Town. Most of the interviewees were entrepreneurs running their own businesses, and almost all were migrants to the city. The overwhelming majority were from other African countries, while one or two had moved to the city from the more rural provinces of South Africa. The numbers of men and women were roughly equal. As a team of three researchers working at different times over a three-year period from 2011–2014, our interviews with these traders were informal and semi-structured. Most participants were interviewed once during this time, in sessions lasting from 20 minutes to over an hour; on a few occasions a second or third conversation was possible. As several participants expressed, early on in the research, their reluctance to have their voices recorded, detailed notes were made during interviews and written up after each day's work. The stories quoted here are therefore not verbatim, but have been reconstructed from these field notes. Most participants were happy, however, for their photograph to be taken, and gave permission for these to be reproduced in published research.

All participants were asked about where they were from, how they had come to be earning a living through beadwork, what they had studied or worked at before entering the craft industry, how they had learned to bead, and how they felt about the work – whether they saw it as art, craft, or something else, and what they saw as the benefits of this kind of work. Most took these questions as cues to talk about their experiences of and feelings about craft rather than the economic impacts of this kind of work. Through their responses and through observation of their everyday production activities, it became clear that the practices of learning to make beaded objects and the act of making itself, as well as the processes of setting up a business and selling their handmade objects, produced multiple forms of personal and social meaning and connection.

For most of these producers, as migrants to the country or at least to the city, craft, with its low start-up costs and relatively easy market access, was a livelihood strategy born of necessity. Their stories emphasised, however, that learning to bead, and making and selling their work, exceeded the economic realm, as personally and socially meaningful experiences that created and sustained relationships of all kinds. The completed beadwork was far from the only or even the most important outcome of these connections; rather, the connections themselves were what most of the stories were 'about'.

The stories of beaded objects, as told by their producers, were therefore seldom about prices, income, and value, although these topics did arise, as I will show. Mainly, however, they focused on the capacities of production practices to produce a sense of community and belonging, structured by dynamics of kinship, collegiality and sometimes unexpected forms of solidarity and difference. In addition, many described how production had restored or provided a new sense of self-identity, with practices of beading and selling facilitating greater emotional and psychological strength. In the sections that follow, quotations and vignettes from the beadwork producers' own experiences are used to demonstrate how the practices of producing bright, contemporary crafts also produce them as people, connected across space and time to both strangers and loved ones.

Producing Individuals

While most of our conversations with craft producers emphasised the social connections generated through their work (and indeed these are the focus in most of what follows) it was clear from several interviews that craft production also contributed to shaping them as individuals. The economic autonomy that this work enabled was a significant narrative strand within producers' accounts of their lives, as Mama Lucy made clear, using words that were resonant across the interviews:

> It is good to know that I can work when I want, but if I don't work hard then I do not get any gain. But I know that when I work hard, all the money that I make is from my work, and I get paid for my hardship … I get out what I put in.

For Boris, who had lost a leg in an accident and was unable to find work, such an achievement was even more powerful. To pass the time, Boris had begun to make beaded wire art in the styles that his grandmother had taught him years before. His wife, Bernice, related their experience:

> My husband realised his days got better when he beaded and so it became a normal occurrence. The number of his beaded items grew and soon I decided we needed to clear it all out to make space. I saw it as a hobby at that time. A friend told me that maybe I could sell the pieces to make some money … I was shocked that someone would pay so much for it! The sale made a good impact on our family as my husband now felt that he was contributing to the family income and this brightened his mood each day. He felt important and useful again.

Like Mama Lucy and Boris, many interviewees emphasised the hard work and perseverance that had been required to develop their skills, stressing that learning their craft had been difficult and even painful at times, but that ultimately they had succeeded. Their stories foregrounded not their economic marginalisation or victimisation, their struggles and hardships, but rather the control they enjoyed over their own lives and the value of their work. These accounts certainly lend themselves to an interpretation of agency, and the temptation to do so, to counter the powerful construction of Southern producers as exploited by the global economic system and its inequalities, is undeniable. However, other producers' narratives hinted that the real meaning of such autonomy exceeded dominant, developmentalist frameworks of agency or choice. That is, the impacts of production for them both ran deeper and were more ordinary and mundane that simply an expanded capacity to act in the world or to change their lives – 'to make their own histories and geographies in conditions of their own choosing' (Lee, 2010: 273). Indeed, even though such full agency or choice were not always available to participants, much value was still to be found in their economic and artistic autonomy.

The artistry of their work was certainly foregrounded by many participants, who described how the act of making had turned them into (or reinforced their sense of themselves as) skilled artisans. This provided them with a status that did not necessarily yield significant economic benefits or the power to bring about life changes, but carried personal and cultural value in its own right. They insisted on the intrinsic value of the hand-mind connection, what might be called 'the intelligent hand' (Sennett, 2008), demanded by their work. As Macy argued, handmade produce should always be seen as art, as they embody both careful thinking and mechanical skill. These producers' craft required 'a lot of imagination' (Alice), 'so much thought' (Macy), constant creativity (Sylvia), and planning and design. These mental processes, coordinated with the body – the arms, wrists, and fingers – in painstakingly developed techniques, constituted the artistic value of the goods they produced and their own status as skilled artisans. As Sifiso suggested, the work demanded by art and craft was something fundamentally human:

> In beadwork, you use your hands and your head to make beautiful things. Both art and craft show the skill which only a person can hold, and so they show what can be made completely by a person.

To reduce this gradual production of skill, self-esteem, pride and emotional engagement, and the pursuance of high-quality artisanal work for its own sake, to notions simply of agency or empowerment, is to miss the emotional and psychological sense of self – the very humanity – that is being honed as part and parcel of pursuing high-quality work and thinking of oneself in those terms. The production of an artist, these producers suggested, may be both more and less than the enhancement of one's ability to bring about change in the world.

Materialising Memories

For many of our participants, craft work enabled a connection to the past through memories of learning to bead. Several of the women, for example, spoke in nostalgic terms about how beadwork had connected them with female friends and shaped a sense of feminine identity and feminine community. Mama Toni, Maybele and Mama Bonni all recalled how learning to bead had provided an excuse to socialise while simultaneously creating a distinctively feminine space in which women's roles could be learned and carried out. As Mama Toni related:

> It is very popular for women in the Congo to bead. We often sit for the whole afternoon under a tree and talk while we are beading. It is a very social activity. Beading for women is a way to leave the house and get away from everyday duties. It is also seen as a very respectable thing to do and so the men do not mind if we sit and bead all afternoon.

Similarly, Maybele reflected on her experience of growing up in rural Mpumalanga, where beading created a distinctively feminine space of learning and leisure:

> We used to often sit with our friends and chat and bead together. It was fun. It was a thing for the women. The men were out all day with the cattle and they learnt different things from each other.

And Mama Bonni, from Kenya, related experiences that were resonant with both of these:

> It seems that I always knew how to bead. Everyone in my family knew how to bead as it was seen as a good social thing to do. All the women would go and sit outside for a few hours and bead while they talked to each other. The children would then also come out and would either sleep if they were still young, or we would play games close to our mothers.

This memory work involved in creative production connected several crafters with loved ones who were distant or permanently absent from their lives. For Lise, for example, beadwork enabled a kind of 'psychic travel' by which she could retreat into memories of her life in Gabon, which she had not wanted to leave. The act of beading provided a way back to relationships and places that were deeply missed, making tangible the memories that helped to make her current circumstances more tolerable.

Having learnt to bead with her friends while growing up, Lise subsequently travelled to Cape Town and gained a degree in electrical engineering. As her husband needed help in his shop, she did not seek work in her field but joined him instead. 'I enjoy beading', she told us, 'It allows me to play with different colours and shapes and sizes. But most importantly it reminds me of my friends back in Gabon and how we used to all bead together'. Lise had not wanted to migrate as all her family were still in Gabon. 'But sometimes it happens that you have to do things that you do not really like to do, like move to another country that you know nothing about', she said.

Fungai, a young man whose stall was made up mainly of beaded animals of all sizes, told a similar story of how beadwork helped him feel both temporally and spatially connected to his family who had remained in Zimbabwe. He expressed his pride in being able to continue the 'family business' of craft production. He and his father shared the work by each spending one month in Zimbabwe making craft with the rest of the family, and the next in Cape Town selling their goods at a large craft market.

> My father, and earlier my grandfather, both made beaded animals and statues using wire to give the statues shape. I was taught by my father and grandfather. I enjoyed it very much and when I finished school I was happy to join the family business. I am very proud to be working with my father and doing the same things my grandfather also used to do.

Like Fungai, Andrew's beadwork materialised memories that connected him with older generations. He related how his grandmother had taught him to bead at home in Malawi. He spoke in detail about their close relationship after the death of his grandfather when Andrew was very young. With her eyesight failing, his grandmother asked Andrew to help her with the beadwork she loved to make. She taught him the skills, telling him she hoped that he would enjoy it as much as she did. After she passed away in 2008, Andrew moved to Cape Town and used his craft skills to support himself.

> My grandmother taught me to bead in many different styles. She showed me how to use wire to make certain shapes and structures. I am now able to make necklaces, earrings and beadwork coasters. My grandmother showed me everything there is to know about the art, how to tie the thread and the wires so it looks neat and how to match different bead sizes and colours. All the things that I bead now are done in the style that my grandmother taught me, as this makes me feel closer to her.

For Lise, Andrew and Fungai, craft production served as a material act of remembering loved ones who were not present, and became a way of connecting with them. The act of making invoked both memories of those relatives and the kin relationship itself, thus sustaining a sense of a community that was no longer manifest in any easily accessible form. Making beadwork provided for these producers a way of making real their place in the world and their belonging to others both seen and unseen.

Shaping Creative Communities

More immediate connections were created across distance and national boundaries for some producers, through a familial division of labour. For entrepreneurs such as Elisa from Kenya, Sifiso from Durban (about two days' bus travel from Cape Town), and Veronica and Mavis from Zimbabwe, the mobilities of beadwork supplies and completed products helped to sustain and develop family networks. All these women had immediate and extended families in their home regions who collected and/or made craft products to send to Cape Town to be sold.

Elisa, for example, travelled home to Kenya every two months to see her family, including her children, and to collect the goods they had accumulated in her absence. Their making and purchasing craft for her 'bought' her free time: 'So when I am in Kenya I do not have to run around and look for goods. I can spend most of my time with my children'.

Sifiso's family in Durban also bought craft products and sent them to her in Cape Town, along with big boxes of beaded necklaces and earrings that they had made. Sifiso herself also made many of the beaded items that she sold, and she proudly declared that her family was the sole producer of beaded jewellery for her stall. And like Sifiso, Veronica and Mavis made their own goods and sold both these and the beadwork that their families at home in Zimbabwe made for their stalls. They often travelled back to Zimbabwe, too, in order to bring back more goods. In Mavis' case, her 'suppliers' included children in the family as young as eight, as well as relatives by marriage, and the profits from the stall were shared equitably among everyone.

For all four of these women, beaded objects were the means by which kin communities were sustained across sometimes great distances. The shared industry of the separated family members cemented their connection through daily activities of buying, making and selling, and the completed boxes of crafts, waiting for collection, both created the need for, and enabled, mobility and face-to-face visits on a regular basis. The mobility of craft objects in production and in networks of trade, play an important role in sustaining transnational lives and networks. With the processes of production and exchange able to be split across nations, yet contained within kin relationships, intimate connections are kept strong even as entrepreneurs leave their families and travel to live and work in distant cities.

Figure 2.2 Beaded necklaces at a craft stall
Source: S. Daya

Connecting Strangers and Defining Others

The material objects and practices of craft production and trade also help to produce new communities, as well as to shape boundaries of inclusion and exclusion. For all participants, beadwork production served to bring together strangers as teachers and students, as colleagues, and/or as traders and tourist-consumers. Isabel, for example, started working at her uncle's stall based in a large craft market 'shed' in Cape Town. She made several friends in the market and some of them taught her to bead. Although she already knew 'the basics', as she put it, her friends taught her more intricate styles.

Mary, similarly, quickly made friends with other stall-holders when she set up a craft stall at a different market to the one where she had started out:

> I particularly became friendly with one man who promised to teach me how to bead. He showed me which beads to buy and which colours to match together. He also showed me the different wires I could bead on and the various beading techniques that were available as well.

Many crafters gave similar accounts of trading, making new friends, and learning new skills from them. Cared, a young Zimbabwean entrepreneur, took pleasure in coming to the market every day and seeing the same people at their stalls: 'They become your family and very close friends'. Macy expressed the same sentiment: 'It is much better now that there are many people around me who I can talk to. I have made many friends'.

Through the processes of learning new production skills and the necessity of sharing spaces of trade, relationships were forged that were often mutually beneficial, as the 'student' typically helped the 'teacher' by working at, and producing stock for, their stall. These were links born of pragmatism and opportunity rather than an initial emotional 'spark', but they were no less meaningful for that. Their situatedness in and around the objects of beads, wire and string, and in the specific geographies of market stalls and squares, highlighted the powerful role of material production in generating the communities that enabled migrants to get by in unfamiliar and sometimes unfriendly, and recently violently xenophobic, cities.

These cheerful accounts were tempered by more ambivalent suggestions of underlying tensions regarding migrants within the informal creative community. While not overt, indications of national or ethnic differences were not entirely absent from the interviews. This was indicated partly in a positive sense; it was evident that migrants' community bonds broadly aligned with national allegiances – Zimbabweans and Malawians, for example, associating mainly with other Zimbabwean and Malawian nationals.

A sense of difference was indicated also through silences. Non-South African craft producers were in fact often reluctant to discuss their experiences of migration and citizenship or their sense of belonging in South Africa, a cautiousness that itself seemed to indicate a perceived hostility. In fact, some participants explicitly expressed a wariness of immigration officials posing as 'researchers'. This sense of perceived animosity at least from law enforcement officials was reinforced by the many references by migrant traders to 'harassment', 'trouble', or 'getting grief' from police officers for trading illegally.

The clearest sense of Otherness within the producer community was, however, expressed by a South African crafter from Zululand, Rasta. Explaining how he had asked someone to teach him to bead, he said: 'but these Zimbabweans are not very forthcoming and did not want to show me how to make them so I had to teach myself'. Later in the conversation he referred again to 'these Zimbabwean guys' who, he claimed, 'distorted' facts about trading permits and copied each other's work.

Similarly mixed views were evident in accounts of producer-consumer connections. These, even more than the collegial relationships with other stall-holders, are restricted within the bounds of commercial transactions and the limited amount of interaction that leads up to and follows on from them. Such forms of sociality both are material and *matter* in an existential sense, giving a sense of worth to the labour and creativity of the entrepreneurs. Participants frequently commented, for example, on the pleasure they felt when consumers complimented their work. Both the admiration from tourists and the pride of the traders in these exchanges were rooted in the same thing: the hand-produced nature of the object. The knowledge of the mental and physical effort required to shape the item in question gave it a value that far exceeded the monetary cost of the raw materials. Mama Lucy explained: 'The tourists often like to buy my bracelets because they like hearing that I myself made them'. And Sylvia agreed: 'I think that beaded items are what the customer wants more because they appreciate the skill and technique that goes into them'. Sifiso, too, emphasised the value of the handmade: 'Tourists really like my necklaces as they always find it shocking that I myself made something like that'. Veronica summed up their sentiments:

> I am very proud when they buy my own beaded products. Since it is so hard to make them, it feels good when you can sell it to someone who really appreciates the work.

But it was clear that not all encounters with tourists were positive. Several participants indicated that the processes of selling also produced less genial forms of connection. For them, these drew lines of identity and difference between consumers in a perhaps unexpected way, marking out locals as sympathetic and

Figure 2.3 Traditional beaded necklace
Source: S. Daya

foreign tourists as (at times) insensitive and ignorant of the realities of life in South Africa. No one among our interviewees proposed the opposite, suggesting that this may be a generally accepted notion of the difference between local and foreign consumers.

While on the one hand traders felt that, as Katrine put it, 'it is normal for us to haggle about the price – it makes the day more fun', it was clear, on the other hand, that the negotiation was often taken too far by foreign tourists, or undertaken in ways that traders considered impolite. As Joselina put it:

> The tourists [as opposed to the locals] have been told that you must try to cut down the price as much as possible. And so when they come here they will start to barter with the price of R5 [about US60c]. What is R5? I cannot feed myself and my husband with R5!

Maybele shared Joselina's displeasure:

> The [foreign] tourists are an interesting group. They often like to come and barter the goods in order to decrease the price. Often this is done in a friendly, good-natured manner. However, sometimes the [foreign] tourists can be quite rough in their mannerisms.

By contrast, there were several remarks by crafters as to the shared understanding of the value of their craft between themselves and local (South African) consumers. Joselina expressed it thus: 'The locals understand us. They will look at what I have to offer, remark on how well it is made and then ask for the price. If they want to buy it, then they will'. Irene articulated the comparison in the clearest terms:

> I also get many people who come to my stall who are locals. I enjoy serving them as I feel they have more knowledge of my products as they understand the cultural meanings that goes into them. They locals are also keen to barter but they do so in a much friendlier way than the tourists. The tourists, it feels, were told they must bargain a price down as low as possible, sometimes so low that it does not even cover the materials it was made from. They also often get angry when you tell them you are not happy with the price. The locals are different. They understand that you also need to make a living and so they are more willing to pay a more just price for them.

As the catalyst for many new, if transient, social interactions, beadwork produces both shared identities and difference. Through learning in close proximity with a fellow trader, and selling to different types of consumers, the everyday production and trading practices of these entrepreneurs create connections that may be momentary or lasting, but in a real sense also produce the people involved: as skilled craft producers, as people who belong in a place, as 'those who are like us' and those Others who are not. Simply put, the processes of making and selling help to materialise communities by shaping boundaries of belonging, including some and excluding others. These accounts demonstrate the fundamentally relational nature of production and its role in generating meaning through everyday material interactions and performances.

Conclusion

In Cape Town, beadwork production is ubiquitous and unremarkable, even mundane. Through that mundanity, however, identities, communities and differences are produced in meaningful ways. Production, especially creative production, these stories demonstrate, is not only about economic imperatives. The lives and livelihoods of Southern producers are not only about financial struggle, economic agency or even uplift and empowerment. Rather, the material practices of production and of trade can shape one's sense of self as a skilled artist, and connect people to each other through learning and friendship, while also marking out differences and establishing boundaries that help to make sense of new places.

These processes are both utterly ordinary and deeply meaningful. The stories recounted here are not exceptional – these kinds of memories, friendships, travels and daily experiences are common to millions of people in South Africa's cities. Our participants' experiences of sustaining connections, forming relationships, developing a sense of self, and getting through the working day are much like everyone else's. But what their stories make clear is that although North-South economic power relations, manifested in the global tourist market, undoubtedly shape these producers' lives, they do not determine all that is interesting about these individuals.

As geography's materialist turn continues to give rise to new theories of things and their implications, it is imperative that production be re-valued as a site of social and cultural meaning. Appreciating the very ordinariness of the identities and communities that are shaped through production practices, as explored in this chapter, should be the aim of this change of perspective. For understandings of production in the global South in particular, this is doubly important if we are to move beyond the dominant narratives of Southern producers that tend towards 'caricature, hyperbole, stereotypes and moralistic hogwash' (Edjabe and Pieterse, 2010: 5). It is only a recognition of this 'ordinary richness' of Southern producers' lives that will enable us to redress the paucity of representations of Southern producers in the geographical literature as rounded people, simultaneously unexceptional and interesting in their own right.

References

Anderson, B. and Wylie, J. (2009) On geography and materiality. *Environment and Planning A* 41, 318–335.

Anderson, B. and Harrison, P. (2010) The promise of non-representational theories, in Anderson, B. and Harrison, P. (ed.) *Taking-Place: Non-representational Theories and Geography*. Farnham: Ashgate, pp. 1–36.

Appadurai, A. (ed.) (1986) *The Social Life of Things: Commodities in Cultural Perspective*. Cambridge: Cambridge University Press.

Barnett, C., Cloke, P., Clarke, N. and Malpass, A. (2005) Consuming ethics: articulating the subjects and spaces of ethical consumption. *Antipode* 37(1), 23–45.

Benson, P. and Fischer, E.F. (2007) Broccoli and desire. *Antipode* 39(5), 800–827.

Castree, N. (2004) The geographical lives of commodities: problems of analysis and critique. *Social and Cultural Geography* 5(1), 21–35.

Colls, R. (2006) Outsize/outside: bodily bignesses and the emotional experiences of British women shopping for clothes. *Gender Place and Culture* 13(5), 529–545.

Cook, I. and Tolia-Kelly, D. (2010) Material geographies, in Hicks, D. and Beaudry, M. (eds) *Oxford Handbook of Material Culture Studies*. Oxford: Oxford University Press, pp. 99–122.

Cook, I. et al. (2004) Follow the thing: papaya. *Antipode* 36(4), 642–664.

Crang, P. (2005) The geographies of material culture, in Cloke, P., Crang, P. and Goodwin, M. (eds) *Introducing Human Geographies*. London: Arnold, pp. 168–181.

Crewe, L. (2008) Ugly beautiful: counting the cost of the global fashion industry. *Geography* 93(1), 25–33.

Daniels, I.M. (2010) *The Japanese House: Material Culture in the Modern Home*. Oxford: Berg.

Dolan, C.S. (2010) Virtual moralities: the mainstreaming of Fairtrade in Kenyan tea fields. *Geoforum* 41, 33–43.

Edjabe, N. and Pieterse, E. (2010) Preface, in Chimurenga and African Centre for Cities (eds) *African Cities Reader*. Vlaeberg: Chimurenga and African Centre for Cities, p. 5.

Green, L. (2008) The social lives of handmade things: configuring value in post-apartheid South Africa. *Social Dynamics* 34(2), 174–185.

Goodman, M. (2004) Reading fair trade: political ecological imaginary and the moral economy of fair trade foods. *Political Geography* 23, 891–915.

Hartwick, E. (2000) Towards a geographical politics of consumption. *Environment and Planning A* 32, 1177–1192.

Hayes-Conroy, A. and Hayes-Conroy, J. (2008) Taking back taste: feminism, food and visceral politics. *Gender Place and Culture* 15(5), 461–473.

Hughes, A. (2000) Retailers, knowledges and changing commodity networks: the case of the cut flower trade. *Geoforum* 31, 175–190.

Jackson, P. (1999) Commodity cultures: the traffic in things. *Transactions of the Institute of British Geographers* 24(1), 95–10.

Jackson, P. (2000) Rematerializing social and cultural geography. *Social and Cultural Geography* 1(1), 9–14.

Jackson, P. (2010) Food stories: consumption in an age of anxiety. *Cultural Geographies* 17(2), 147–165.
Kearnes, M.B. (2003) Geographies that matter – the rhetorical deployment of physicality? *Social and Cultural Geography* 4(2), 139–152.
Kirsch, S. (2012) Cultural geography I: materialist turns. *Progress in Human Geography* 37(3), 433–441.
Lee, R. (2010) Spiders, bees or architects? Imagination and the radical immanence of alternatives/diversity for political-economic geographies, in Fuller, D., Jonas, A. and Lee, R. (eds) *Interrogating Alterity: Alternative Economic and Political Spaces*. Farnham: Ashgate, pp. 273–288.
Lees, L. (2002) Rematerializing geography: the 'new' urban geography. *Progress in Human Geography* 26(1), 101–112.
Leslie, D. and Reimer, S. (1999) Spatializing commodity chains. *Progress in Human Geography* 23(3), 401–420.
Mansfield, B. (2003) 'Imitation crab' and the material culture of commodity production. *Cultural Geographies* 10, 176–195.
Miller, D. (1998) Coca-cola: a black sweet drink from Trinidad, in Miller, D. (ed.) *Material Cultures: Why Some Things Matter*. Chicago: University of Chicago Press, pp. 169–188.
Miller, D. (ed.) (2001) *Home Possessions: Material Culture Behind Closed Doors*. Oxford: Berg.
Miller, D. (2004) Making love in supermarkets, in Amin, A. and Thrift, N. (eds) *The Blackwell Cultural Economy Reader*. Oxford: Blackwell, pp. 251–265.
Miller, D. (2010) *Stuff*. Cambridge: Polity.
O'Neill, T. (1999) The lives of the Tibetonepalese carpet. *Journal of Material Culture* 4(1), 21–38.
Pfaff, J. (2010) A mobile phone: mobility, materiality and everyday Swahili trading practices. *Cultural Geographies* 17(3), 341–357.
Ramsay, N. (2009) Taking-place: refracted enchantment and the habitual spaces of the tourist souvenir. *Social and Cultural Geography* 10(2), 197–217.
Scheld, S (2003) The city in a shoe: redefining urban Africa through Sebago footwear consumption. *City and Society* XV(I), 109–130.
Sennett, R. (2008) *The Craftsman*. London: Penguin.
Tiffany, D. (2001) Lyric substance: on riddles, materialism and poetic obscurity. *Critical Inquiry* 28(1), 72–98.
Tolia-Kelly, D. (2004) Materializing post-colonial geographies: examining the textural landscapes of migration in the South Asian home. *Geoforum* 35, 675–688.
Whatmore, S. (2006) Materialist returns: practising cultural geography in and for a more-than-human world. *Cultural Geographies* 13,600–609.
Woodward, S. (2007) *Why Women Wear What They Wear* Oxford: Berg.
Yeh, E.T. and Lama, K.T. (2013) Following the caterpillar fungus: nature, commodity chains, and the place of Tibet in China's uneven geographies. *Social and Cultural Geography* 14(3), 318–340.

Chapter 3
People, Place and Fish: Exploring the Cultural Ecosystem Services of Inshore Fishing through Photography

Tim G. Acott and Julie Urquhart

Introduction

The Millennium Ecosystem Assessment (MEA, 2005) set out a framework for understanding the benefits that humans derive from the environment in order to inform decision making. It categorized these benefits as: *provisioning services,* such as food, water, timber; *regulating services*, such as climate control, waste, water quality; *supporting services,* such as soil formation, photosynthesis, nutrient cycling; and *cultural services,* such as recreational, spiritual and aesthetic benefits. Since then there has been a plethora of research and wider interest in devising ways of assessing and measuring those services (Haines-Young and Potschin, 2009; Sagoff, 2011; Shan and Swinton, 2011), often involving economic valuation techniques devised by economists and ecologists. While these can be useful for assessing the provisioning, supporting and regulating services, measuring or assessing the cultural services that humans receive from ecosystems has proved to be more problematic. However, there is increasing recognition of the role of multiple disciplines in understanding the complex and multi-faceted ways that ecosystems shape culture and cultural value.

There is a growing research agenda on the use of social science approaches to cultural ecosystem services (CES) (Chan et al., 2012; Milcu et al., 2013), however less attention has been paid to the contribution from the arts and the humanities. In this regard, Coates et al. (2014) assert that social science methods (whether quantitative or qualitative) might benefit from drawing on approaches from the arts and humanities when it comes to the consideration of CES. Some of the issues include interviewees being unaware of the existence of CES or the ability of people to articulate or reflect on cultural values (Bieling and Plieninger, 2013). In order to address this, the UK National Ecosystem Assessment Follow-On called for a more explicit integration of the arts and humanities with social science in order to deepen and broaden the discussion of CES (Coates et al., 2014). One way that the arts and humanities can play an important role in adding value to ecosystem services "is for creative practitioners to produce inspiring poems, paintings, films and other artworks, based on a reflective process informed by evidence of the cultural benefits of Ecosystem Services" (Coates et al., 2014, 33).

In this regard, a range of different creative media can be points of engagement between people and the natural world and can encourage people to explore places that are shaped both by nature and culture (Coates et al., 2014). Creative media can be used to represent aspects of the natural world, and can be an important way that people engage with ecosystems. As such, photography, through the representation of different human-environment relationships, can bring new cultural worlds into being and focus attention on issues that might otherwise remain in the background.

It is with the key question of "what is the role of photography for our understanding of CES?" that we approach the subject of this chapter. We, firstly, consider the role of photography in the co-production of culture and how it can be both creative practice and a social science research tool. Secondly, we draw on the experience of two photographic projects, which were conducted as part of European research programmes carried out between 2009–2014, to reflect on the role of photography in understanding CES through an exploration of sense of place in inshore fishing communities.

Creative Practice and the Co-production of Culture

From a social science perspective there are different ways that photography can make a contribution to understanding CES. First, in the discipline of geography it is common to use photography as a means of recording information about natural and built environments. For instance, geological field sections, a particular landscape view or perhaps an architectural detail. Photography can be used to build up a collection of images that record key details of features that contribute to sense of place. Landscape character assessment and urban character assessment make use of photography as a way of documenting key features of interest (for instance the work by Natural England on National Character Areas). The importance here is not the aesthetic quality of the images but the information that is being depicted and cataloguing for subsequent analysis. This experience is consistent with photographic traditions dating back to the nineteenth-century that saw increasing acceptance of the 'authority' of a photograph to show the world as it really is.

However, this authoritative view can be contrasted with a representational perspective, as Sontag (1977) suggests: "Although there is a sense in which the camera does indeed capture reality, not just interpret it, photographs are as much an interpretation of the world as paintings and drawings are" (6–7). Photographs are not wholly objective but neither are they completely subjective (Ward, 2004). Sontag (1977, in Ward 2004) explores the meaning of photography as lying between two poles of beauty (self-expression and concern for emotion and aesthetic) and truth (communication). This continuum aptly captures how photography can traverse the relationship between image and reality with a creative element. Yet there is a difference between photography and other forms of representation, in that a photograph is tied to a material reality in a way that, for instance, a painting is not.

Second, the role of photography in social science research spans many disciplines including anthropology, environmental psychology and human geography (Markwell, 2000). Images can be integrated with other forms of information, even where the central focus is not the analysis of the visual. In this case images and photographs are not just sources of data, they help to facilitate the process of research (Gold, 2004). In this latter sense photography can mediate the relationship between a researcher and the subject: "In retrospect, I realize that I learned as much from the social interactions involved with taking photographs, showing images to respondents, and sharing prints with colleagues and students as I did from analyzing what is shown in the images themselves" (ibid.: 151). This mediation might occur in numerous ways, for instance, seeking permission to photograph somebody can be the starting point in developing a relationship with them. A camera in the field can be the point around which a discussion can begin between researcher and subject. This interaction between research and participant can continue as the photographs are produced and presented to an audience.

Showing photographs to participants can be a form of photo-elicitation (see Harper, 2002) and the starting point for individual and group interviews. Photography can help social science researchers to understand and explore the meanings that environments have for people. Photo elicitation can take many forms (Van Auken et al., 2010; Stewart et al., 2004; Kerstetter and Bricker, 2009) but the key idea is that photographs are used as a starting point to stimulate discussion with individuals or groups about what a place means to them. The photographs can be taken by members of the community or by the researcher. The photographs provide a stimulus for a resulting conversation. As with the use of photographs to record information, the aesthetic quality of the pictures are really of secondary concern, the emphasis is on how the photographs can promote discussion on the meaning that environments have for people. Beyond the individual, photography also has a role to play in community development: "it is becoming clear that community use of photography can be used to give voice to, and make visible, otherwise hidden groups and community-based issues" (Purcell, 2007, 112). In thinking about the process of photography and the creation of cultural value the following section turns to creative practice and the co-production of culture.

In photography for policy related research it is important to consider the practice of taking photographs and the relationship that is co-constructed between the researcher, human participants, the non-human world and policy makers. Understanding the importance of photography is to see the act of taking a picture as a process that brings new worlds into existence as old tropes are challenged and new narratives can be told. As Sontag (1977, 11) argues: "A photograph is not just the result of an encounter between an event and a photographer; picture taking is an event in itself, and one with ever more peremptory rights – to interfere with, to invade, or to ignore whatever is going on. Our very sense of situation is now articulated by the camera's interventions". Crang (1997, 360) describes these linkages as a 'circuit of culture' where he suggests that it is important to examine how cultural products are actually taken up and used: "The circuit elaborates the flows from producers to product to consumers and back in a developing and ever-changing spiral, as each works with the materials of the previous stage". But, as Crang cautions, "this can too easily imply that the consumption practices are a separate field from those of production" (ibid.: 360).

Photography connects the photographer with subject and then to an audience where the photographic representations are circulated. The process of becoming a viewing subject is connected with ways of seeing the world (Rose, 1992 in Crang 1997). The circuit of culture begins to talk to the performative turn that has become increasingly important in geographical research. In this sense all human practice is understood as being 'performed' in a public presentation of the self. Through conceptual work such as 'non-representational theory' (Thrift, 2008) the focus of enquiry has shifted from representation to ideas of performance and practice (Wiley, 2007). Images not only represent a social reality, they also shape the way people think (Burri, 2012). The practice of photography, therefore, is as important as its representations. Practice refers to the production of the photographs as well as their subsequent circulation in society.

Pictures have the capacity to frame ontology through bringing our attention to certain aspects of the world over others, or in Heidegger's terms to make part of the world 'occurrent' (Crang, 1997). This chapter aims to explore the intersection of 'occurrence' with photography, fisheries and policy and explore how photography might be useful for understanding CES in the context of sense of place in inshore fishing communities. Through two research projects[1] spanning five years and covering four European countries photography has been integrated in multiple ways to explore and make visible the cultural services that arise through the practice of inshore fishing in coastal communities. The focus was on how inshore fishing contributes to the creation of a particular sense of place that is important for both residents and visitors in these locations. We set out the range of photographic approaches that were used and reflect on their utility for revealing and, in some cases, producing cultural values associated with inshore fishing.

In doing so we want to move beyond thinking about photography as simply record keeping or representation and consider perspectives that span the social sciences and the arts and humanities with emphasis on both the processual dimensions and the end image. Crossing these disciplinary divides is not an easy task as it entails embracing a range of contested ontologies and epistemologies. For the social scientist important questions might include: What type of data does photography produce?; What is the relationship between data, the researcher and the subject?; What guidelines or approaches should be followed when using photography for research purposes? From an arts perspective emphasis may instead be placed on the creative process and the production of new, visually arresting or

1 CHARM III (co-funded by the INTERREG IVA Channel Programme, 2009–2011) and GIFS (co-funded by the INTERREG IVA 2 Seas Programme, 2011–2014).

meaningful images as well as critical thinking in the arts relating to innovation in technology and style. It is in the synergy between social science and the arts that we feel there is the greatest salience of photography to contribute to broad policy-making and community development agendas in natural resource management.

Through these projects we explore the use of photography as both a creative process and as a tool that can provide meaningful engagement with communities and individuals around the practice of inshore fishing. In this regard, photography is used to both elicit and create CES values. However, in order to engage with people and communities it was important to use terminology that they would recognise and be comfortable with. The UKNEA recognised that most people are more comfortable with terms such as 'nature', 'place' and 'landscape' (which carry greater cultural meaning for people) rather than terms 'ecosystem' or 'ecosystem services' (UKNEA, 2011). In order to capture both the perceptual, experiential and the situated values of people we adopted the idea of 'sense of place' as a conceptual, and yet familiar, framework. The following sections, firstly, give a brief overview of the concept of sense of place, followed by the differing ways that photography has been used in our projects.

Sense of Place

There is increasing interest in using the idea of sense of place in the management of natural resources (Williams and Stewart, 1998; Farnum et al., 2005; Cantrill, 1998; Kianicka et al., 2006) particularly when related to the idea of ecosystem services (MEA, 2005; UKNEA, 2011). There is an abundance of literature on sense of place that spans numerous academic disciplines including humanistic geography, environmental psychology, sociology and architecture (Davenport and Anderson, 2005; Kyle and Chick, 2007; Clay and Olson, 2007). However, many studies make reference to landmark work conducted in the 1970s by humanistic geographers Yi Fu Tuan and Ed Relph (Tuan, 1974; Relph, 1976). They draw on phenomenological perspectives to suggest that sense of place refers to the emotional meanings that people have for places and is grounded into social relationships and processes that occur in particular settings (Acott and Urquhart, 2014). Thus, sense of place is about trying to understand complex human-environment relationships by exploring the meanings that people construct and attribute to places (Kaltenborn, 1998). However, with the emphasis on meaning it is important to remember that sense of place is also grounded in a material physicality and places are defined by their physical environment (Stedman, 2003). Malpas (2008) reminds us that there is a common tendency to view culture as something that is additional to and separate from its materiality. Eisenhauer et al. (2000) assert that there is a reciprocal relationship between physical environments and people in what Crist (2004, 12) calls a "cultivation of receptivity" in which humans can receive meaning from the world through "opening oneself, listening, watching, being within, letting be, or merging into". In this sense, social life and culture will influence place meanings, but the material elements of place are also important.

In our work we were interested in the way that photographic representation is also a form of practice and engagement with the world. Making use of a camera mediates our engagement with the world and the production of images draws in broader participants in a process of 'world making'. In other words, photography used in this research moves into a co-constructionist sphere whereby new networks are created that give rise to particular actants and negotiated ways of knowing. Both projects focused on the CES that arise as a result of inshore fishing along the coasts of the English Channel and Southern North Sea. The aim was to understand the cultural benefits that arise from the activity of inshore fishing by exploring how it contributed to 'sense of place' in coastal towns. The following sections outline how photography was used:

1. As an auditing tool to record and document the physical environment
2. As a tool to help individuals reflect on what is important about a place (exhibitions)
3. As a mediator between researcher and subject to facilitate interviewing (photo-elicitation)
4. As a creative endeavour that creates representations of places and thus contributes to place making (professional photography)

Auditing the Visible CES of Inshore Fishing

The coasts of southern England and northern France are well known for their fishing towns and villages. For example, the numerous coves and inlets of Cornwall, dotted with fishing boats, either moored in picturesque harbours or drawn up onto beaches. Or charming French harbours such as Saint-Vaast-la-Hougue, once a thriving port for the Newfoundland fishing fleets, but now supporting an inshore fleet and oyster and mussel fishery. These fleets of small boats have a particularly important role to play in creating distinctive place identities in small towns and harbours and have resulted in the production of a wide range of material objects, in both past and contemporary practice. During research as part of the CHARM III project in 2009–2011 over 75 coastal towns and villages were visited in England and France and a photographic survey of objects, activities and urbanscapes that related to fishing was completed (Acott and Urquhart, 2012; Urquhart and Acott, 2014; Urquhart and Acott, 2013).

Decisions had to be made about which objects to include in the survey. While some objects were clearly directly related to marine fishing others referenced general maritime activities more broadly. These objects represented ways in which the activity of marine fishing was being translated into cultural artefacts which created tangible objects that contributed to a sense of place and place character within communities. The types of objects were wide ranging but included fishing boats, nets, pots, books, buildings, paintings, tourist souvenirs, information boards, monuments, street furniture and so forth. While some of these objects contributed to character in a clear and obvious way (e.g. the fishing boats) others were less obvious (e.g. a decoration hung in a window, or a fisheries-related door knocker). Nevertheless, all the objects were visible from public places and in that way helped to contribute to the overall character of a place.

Exhibitions

In the GIFS project, the CHARM photo auditing was extended with two principal researchers taking photographs of activities and objects associated with inshore fishing. They visited different towns and locations as outsiders to the fishing industry but took a series of photographs that were tangible, visible evidence of the CES associated with fisheries (Figure 3.1).

The intention here was not to simply record objects associated with inshore fisheries. The resulting photographs would be used in a travelling exhibition (Figure 3.2) visiting seven locations (Looe, Whitstable and Wells-next-the-sea in England, Le Guilvinec, Rennes and Saint-Vaast-la-Hougue in France and Oostende in Belgium) over the summers of 2013 and 2014. The exhibitions were a mechanism for engaging visitors to explore inshore fishing in relation to CES and were organised under themes taken from the Millennium Ecosystem Assessment and included aesthetic values, cultural identity, education and knowledge, heritage values,

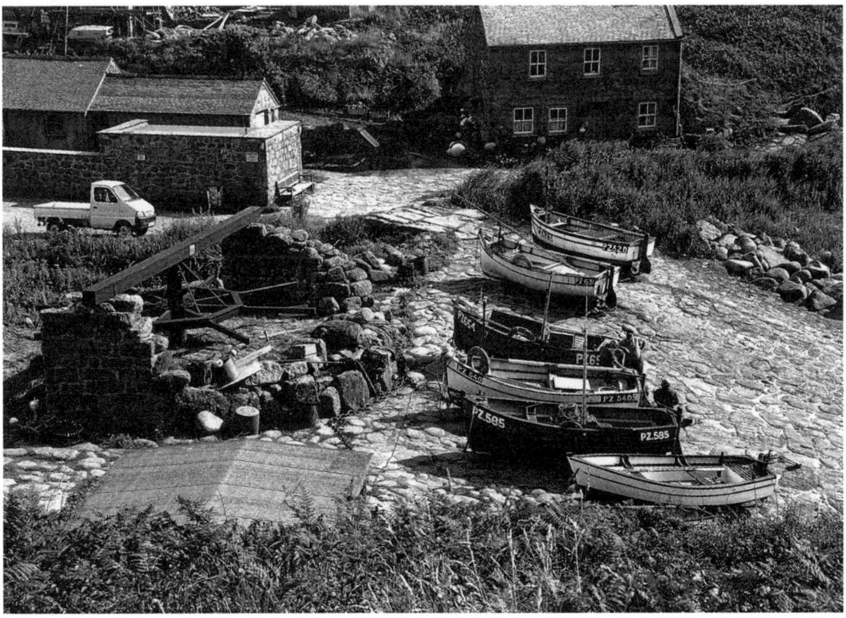

Figure 3.1 Capstan Wheel at Penberth, Cornwall. Visible evidence of cultural heritage
Source: T. Acott and J. Urquhart

Figure 3.2 Community exhibition in Looe, Cornwall
Source: T. Acott and J. Urquhart

inspiration, social relations, spiritual and religious values and tourism and recreation. Each theme had a number of pictures associated with it and a small amount of text that described the theme and provided some context to the picture. In addition to the researcher photographs, people living in the communities were also invited to submit photographs and a short textual description to the exhibitions. The objective of this part of the research was to create a narrative around the importance of inshore fishing and to highlight the many different ways the activity could be valued.

Throughout the course of the exhibitions various interactive elements were introduced in order to test their efficacy for promoting community participation. Initially it proved difficult to get people visiting the exhibitions to write down comments. As the exhibitions progressed efforts were made to develop interactive elements (Urquhart et al., 2014). Statements were designed around a visual five-point Likert scale where members of the public could indicate their views by placing colour-coded stickers of an animated face on the scale (Figure 3.3). The stickers were colour-coded in an attempt to gain demographic information, with yellow stickers representing the views of residents and red stickers representing the views of visitors. In addition there was also a comment box placed by each statement that enabled members of the public to anonymously provide more views relating to the statement if they so wished.

Figure 3.3 An example of a statement and response scale
Source: T. Acott and J. Urquhart

Initial results suggest the exhibitions are providing new ways for people to understand the cultural importance of inshore fishing. Many comments on the quality of the exhibition were provided in the visitors' book, for example:

- "Well done, a great show – a way into seeing anew" – Looe
- "An excellent exhibition, which captures a key element of the life of the town. Images are used to considerable effect and the range of perspectives brought to bear on Whitstable is compelling. Very interesting" – Whitstable
- "Beautiful, thought provoking and important, all strength to you" – Looe
- "I think the interactive aspects of this exhibition are important as participation is a growing element of art where the artist can become the facilitator, so that the public cease to feel disconnected and become more involved in the creative process thus giving it life and new ideas from the outside. The project is then holistic" – Saint Vaast
- "The interactive nature of the display is fantastic and really engaging and appealing to both young and old – fantastic job!" – Whitstable

Submissions by local communities included a range of subject material as illustrated in Table 3.1. The number of different subjects taken helped to make visible the diverse ways inshore fishing contributes to cultural value. By linking to the MEA themes it was possible to deliver a narrative about the way that inshore fishing is relationally associated with a broad variety of terrestrial activities (e.g. heritage, songs, artworks, sculptures, monuments). An awareness of this cultural

Table 3.1 Range of photographs submitted by subject in order of popularity

	France		Belgium		England		TOTAL	
	No.	%	No.	%	No.	%	No.	%
Boats	26	37	1	11	44	31	71	32
Harbour/ seascape	22	31	0	0	28	20	50	23
Fishers/ processors	10	14	2	22	24	17	36	16
People on shore	0	0	1	11	10	7	11	5
Fishing	6	8	1	11	4	3	11	5
Tourism	1	1	1	11	10	7	12	5
Gear	1	1	0	0	7	5	7	3
Fish market	2	3	2	22	2	1	6	3
Seagulls	0	0	0	0	7	5	7	3
Art	1	1	1	11	3	2	5	2
Auction	2	3	0	0	0	0	2	1
Fish	0	0	0	0	3	2	3	1
Signs	0	0	0	0	1	1	1	0

complexity is generally not present in many fisheries related policy developments, although the recent revision of the European Common Fisheries Policy does allude to the importance of small-scale fisheries ((EU) No 1380/2013). Ongoing work is developing the exhibition so that it can be used as a group photo-elicitation methodology. The key element here is to transform the normally passive experience of an exhibition into one where the visiting audience wants to provide reflective feedback about their experience.

Photo-elicitation

Photo-elicitation can help to reveal the importance that people attach to a place (Acott et al., 2014). There are different types of photo-elicitation (Purcell, 2007; Fink, 2011; Holgate et al., 2012; Johnson et al., 2008) but, in the GIFS project, 'researcher-photography' was used as a form of photo-elicitation interview (PEI) to explore the role of inshore fishing in shaping the relationship of people to place. PEI can take many forms but is generally used where photographs facilitate discussion between researcher and interviewee. The rationale is that a series of photographs can be a starting point for a conversation that can evoke deeper reactions than just speaking to someone without visual prompts. Six case studies were undertaken (Wells-next-the-sea, Isle of Wight, Beer and Looe in England; Le Guilvinec in France and Oostenduinkerke in Belgium) involving about 10 participants in each location (Kennard, in prep). A number of photographs were taken by the researcher that depicted issues around exploring the cultural values of inshore fishing (for example Figure 3.4). This approach to PEI gives the researcher control over the photographs used in the discussion and therefore has more ability to direct the conversation. This can be an advantage in that the researcher might be able to introduce ideas and topics that the interviewee had not thought about. Alternatively, participant-elicited photography can be used, where the participant is asked to take photographs that depict what is meaningful to them.

A case study of the PEI approach used in Oostduinkerke, Belgium, provides an example of how the cultural values of residents were explored (Acott et al., 2014). Oostduinkerke is the location of horseback shrimp fishermen (Paardenvissers). This is a non-commercial fishing operation now supported by the tourism industry but which has recently been given World Heritage Status. The importance of the fishing to the cultural identity, heritage values, spiritual services and recreation / tourism were clearly recognised in the PEI study as illustrated by the following quotes from the interview transcripts (Acott et al., 2014; Kennard, in prep):

- "It's something that is important that I want to cherish and safeguard ... it's the beating heart of Oostduinkerke" and " ... it lives among the people" (82).
- "In order to have a future for the fishermen we have to look at the past, and learn from it" (82).

- "These are people that spend their entire time at the beach – they feel very connected to the sea and the fishing on horseback is a passion of theirs. They want to be connected to the sea on a daily basis. They are people who cannot live without the sea" (83).
- "The fishing used to be more important than tourism, but now tourism has become more important than the fishermen" (83).

Figure 3.4 Horseback fishing in Oostenduinkerke, Belgium. CES themes include education, tourism and heritage
Source: M. Kennard

Professional photography

In the GIFS project the use of photography was extended by hiring a professional photographer, Vince Bevan, to produce a photo-documentary of fishing places in case study locations along the English Channel and Southern North Sea. Vince is an experienced photojournalist with work published in, amongst other editorials, *The Guardian Weekend* and *Geographical Magazine*. His assignments have taken him to many parts of the world including Bosnia and East Timor. The brief was to explore the 'landscapes of fishing' in different parts of the study area. He was asked to capture both the diverse landscapes that he encountered, but also the way that fishing activity was visible in the environments that he visited. The result of his work is a stunning collection of online images and a series of national photography exhibitions starting at the National Maritime Museum in Falmouth (29 March to 18 May 2014) and then travelling to Belgium and the Netherlands throughout the summer of 2014.

The intention of this part of the project was to create a series of visually arresting images that would cause people to take notice and reflect on the issues being depicted (Figure 3.5). In this case photographs were not being used to categorise (unlike the auditing) but a more creative approach was encouraged that made use of the professional photographer's skill and artistry. For instance, the use of saturated colour and light to add drama, the use of shutter speed and aperture to create blur and differential areas of focus. However, this more creative approach did result in various discussions during the course of the project. For instance, initially there was a concern that the images being created were overly romantic depicting a somewhat stereotypical image of pretty boats. Of course, the harsh reality of inshore fishing is very different from this image, with death and injury a constant spectre for many fishing families. From the outset, it was the intention of this part of the project to tell a story around inshore fishing in the early twenty-first century that would resonate with a broad variety of audiences. In this sense then, this activity was in part recording images for posterity. But in another sense, it was creating a narrative that sought to get people to reflect and think about the diversity of inshore fishing activity.

These examples of the types of photography used in two research projects start to illustrate photography as a co-constructed activity creating relational associations between the photographer, the audience and the place. It is a tool to record what is in the environment but is also a creative practice around which new narratives can be constructed. Photography facilitates relationships between researchers and their subjects while also creating new networks of social exchange as pictures are displayed in exhibitions or circulated on the internet.

Figure 3.5 Herring Festival, Boulogne-sur-Mer
Source: Vince Bevan, GIFS Project

Discussion

Photography can play numerous roles in mediating the relationship between people and ecosystems and can contribute to the creation and recognition of environmental values through the development of new networks and sharing of knowledge and information. The following sections discuss the lessons learned from the photography deployed as part of the CHARM III and GIFS projects and argues that photography has an important role in developing policy related perspectives for understanding sense of place and cultural ecosystem services.

Photography was used to document and catalogue phenomena that represented cultural value of inshore fisheries as captured in tangible objects. This approach is consistent with traditions of photography in the nineteenth century that saw governments, the military and commercial organisations turn to photography as the medium most able to record the world accurately (Wells, 2011). Photography has since been employed in descriptive and analytical ways in many social science disciplines including sociology and geography. However, with its focus on a realist ontology that can be accurately depicted this use of photography pays more attention to product and less attention to process. In the case of the GIFS project, using a camera to record those cultural objects that are associated with inshore fishing provided the researcher with a tool that could help guide their observation. Suchar (1997) develops this point and talks about the 'interrogatory principle' of photography. For Suchar the documentary potential of photography is not inherent in the photographs but in the interactive process when photographs are used to explore a particular subject. In our work using a camera helped draw out and make visible background objects that contributed to a sense of place, yet perhaps remained unnoticed for many people at that location. For instance, people using fishing-themed decorations on their houses, benches adorned as memorials to fishermen, tourist wares vying for attention in shop windows.

In our work photography as a process facilitated relationship building between researchers and subjects. This was through the use of exhibitions and PEI techniques with photographs being used as a starting point for conversations about inshore fishing. However, just the act of carrying a camera around and taking photographs of unusual subjects (e.g. signs with fish motifs) could be enough to create a new conversation. Also the importance of the camera to slow the researcher down so that care and attention to detail are considered should not be underestimated. In the case of inshore fishing our photographs were not just representations of reality, they were the starting point for creating a new awareness and understanding of inshore fisheries. In the words of Crang (1997, 362): "Images are not something that appear over and against reality, but parts of practices through which people work to establish realities. Rather than look to mirroring as a root metaphor, technologies of seeing form ways of grasping the world". In our case we were trying to demonstrate the many relational associations that inshore fishing contributes to communities and in so doing were actively promoting cultural value (creating cultural value as opposed to eliciting cultural value).

Part of the process of photography is the creation of a picture, either as physical print or a digital image. We used these representations to develop a narrative around which the values of ecosystems are considered. Through attendance at our exhibitions we presented a way of thinking about ecosystem value that was unfamiliar to many people but consistent with the idea of CES as described in the MEA. In developing such exhibitions photography is moving out from the social science framework of an ontologically realist tool for recording, into the creative sphere of the arts and humanities with a focus on representation and the creation of narratives. The salience of this aspect of the GIFS work, and potentially further afield, should not be underestimated. As articulated in the UKNEA Follow On:

> Creative approaches influenced by research in the arts and humanities not only provide new forms of evidence for decision-makers, but can help engage communities and engender stewardship of local natural resources; such approaches may be particularly effective when incorporated into a learning curriculum, for instance. Linking these techniques to wider tools and approaches developed in the landscape and heritage sector represents an opportunity for future innovations in the practical application of cultural ecosystem services concepts. (Church et al., 2014, 6)

The experiences of the GIFS community exhibitions were interesting in that initial feedback indicated that many people had not really thought about the broader cultural contribution that inshore fisheries make to coastal towns. Presenting the photographs in the form of a narrative aligned to cultural ecosystem services helped to put a focus on the importance of fisheries for identity and sense of place in coastal communities. This finding echoes Bieling and Plieninger's (2013) concern that a lack of awareness amongst people may hinder normal CES elicitation techniques.

Social science can help elicit ecosystem values from people, whereas the arts and humanities also have a role in shaping new meanings and creating new value (Coates et al., 2014). In the context of GIFS the professional and community exhibitions are already highlighting the cultural ecosystem services of inshore fisheries. This blurring of research / creative output is reflected by Smith (2014, 4): "Much art is about the experience of the moment, whereas most research is about recording or analysing something after an event". Photography can play a role in both of these aspects. Photography can help capture, document and analyse but it also provides a vehicle for expressing emotive and aesthetic themes to be communicated to wider audiences. In sense of place research the GIFS photographic element brought to the fore the idea that photography is both researching the identity and heritage values of a place, but at the same time it is also contributing to the creation of those values. Perhaps in the same way that books like *Edgelands* by Roberts and Farley (2012) can draw attention to the cultural values of unfamiliar and unacknowledged nature, photography can be used to highlight and communicate how nature can be translated into a myriad of cultural values. In the case of inshore fishing photography can offer momentary glimpses into the dangerous world of the last hunters.

Photography also has a role in taking manifestations of place-based values away from their immediate locale and transporting them to far-away places. The professional photography undertaken as part of GIFS has taken the photographs from coastal towns and villages and displayed them in venues in England, the Netherlands and Belgium. The photographic exhibition is a conduit through which the distant valuation of ecosystems can take place. The photographs are removing the need for a direct experience of place. This is not just an academic point. This valuing at a distance can stimulate public support for activities such as inshore fishing and has the potential to be translated into new economic opportunities through the development of responsible tourism initiatives.

Conclusion

We have explored the idea of photography as a co-constructed activity that connects the researcher (photographer), the subject being photographed and the viewing audience. These relational associations embedded in ideas of sense of place and CES begin to challenge the idea of objective research. By taking documentary photographs we were developing a narrative of place, in Heidegger's terms we were making occurrent the importance of the contribution of inshore fisheries to sense of place. This sense of place narrative was also used to explore the idea of CES and in doing so create and disseminate cultural value. In our case the research included photographic auditing, photo-elicitation and professional photography. These resulted in the creation of a series of exhibitions that created a representation of inshore fishing that made visible relational associations between the sea and the land that might have otherwise been hidden from the general public and policymakers alike. The results of the process, therefore, break down distinctions between objective science and artistic creation.

Photography is both revealing a world but at the same time is bringing a world into being. Our research has begun to highlight the importance of understanding how the creative process can contribute to the creation and elicitation of cultural value. The policy-making agenda is often focused on the use of numeric empirical data. However, we are suggesting that understanding the relationship between creative processes and cultural values should be incorporated more fully into policy making. For this to happen the use of a broad range of qualitative and quantitative evidence needs to be admitted together with a realisation that cultural value is something that can be generated as a result of creative processes.

At the outset of this chapter we posed the question "what is the role of photography for our understanding of CES?" By way of work carried out as part of two projects, CHARM III and GIFS, we suggest four ways that photography can be used in the development of CES perspectives:

- As a tool for recording the cultural artefacts produced as a result of the use of ecosystems
- As a process to facilitate engagement between researchers and communities
- As an approach to create new cultural values by developing narratives around cultural ecosystem services and sense of place
- As an education tool to raise awareness of the value of natural resources leading to stewardship and a deeper understanding of those resources

It is hoped that these four interrelated elements can be used as a starting point in CES research to highlight that understanding cultural values is not just about eliciting views, it is also about creating new narratives and programmes of communication and education that create new values. In this sense photography, sense of place and CES are co-constructed around networks of associations spun throughout subjective and objective worlds.

Acknowledgements

The projects reported in this article were co-funded by the INTERREG IVA Channel Programme project CHARM III and INTERREG IVA 2 Seas Programme project GIFS (2011–2014).

References

Acott, T. and Urquhart, J. (2012) Marine fisheries and sense of place in coastal communities of the English Channel, Final report prepared as part of the INTERREG IVA CHARM III project, Chatham: University of Greenwich.

Acott, T. and Urquhart, J. (2014) Sense of place and socio-cultural values in fishing communities along the English Channel, in Urquhart, J., Acott, T., Symes, D. and Zhao, M. (eds) *Social Issues in Sustainable Fisheries Management*. Springer: London, 257–278.

Acott, T., Urquhart, J., Church, A., Kennard, M., le Gallic, B., Leplat, M., Lescrawauet, A.-K., Nourry, M., Orchard-Webb, J., Roelofs, M., Ropars, C. and Zhao, M. (2014) *21st Century Catch*. University of Greenwich, Chatham.

Bieling, C. and Plieninger, T. (2013) Recording manifestations of cultural ecosystem services in the landscape. *Landscape Research* (38), 649–667.

Burri, R.V. (2012) Visual rationalities: Towards a sociology of images. *Current Sociology* 60, 45–60.

Cantrill, J.G. (1998) The environmental self and a sense of place: Communication foundations for regional ecosystem management. *Journal of Applied Communication Research* 26, 301–318.

Chan, K.M., Guerry, A.D., Balvanera, P., Klain, S., Satterfield, T., Basurto, X., Bostrom, A., Chuenpagdee, R., Gould, R., Halpern, B.S., Hannahs, N, Levine, J., Norton, B., Ruckelshaus, M., Russell, R., Tam, J. and Woodside, U. (2012) Where are cultural and social in ecosystem services? a framework for constructive engagement. *BioScience* 62, 744–756.

Church, A., Fish, R., Haines-Young, R., Mourato, S., Tratalos, J., Stapleton, L., Willis, C., Coates, P., Gibbons, S., Leyshon, C., Potschin, M., Ravenscroft, N., Sanchis-Guarner, R., Winter, M. and Kenter, J. (2014) UK National Ecosystem Assessment Follow-on. Work Package Report 5: Cultural ecosystem services and indicators, UNEP-WCMC, LWEC, UK.

Clay, P. and Olson, J. (2007) Defining fishing communities: Issues in theory and practice. *NAPA Bulletin* 28, 27–42.

Coates, P., Brady, E., Church, A., Cowell, B., Daniels, S., DeSilvey, C., Fish, R., Holyoak, V., Horell, D., Mackey, S., Pite, R., Stibbe, A. and Waters, R. (2014) Arts and Humanities Working Group (AHWG): Final Report, UK National Ecosystem Assessment Follow-on, UNEP-WCMC, LWEC, UK.

Crang, M. (1997) Picturing practices: research through the tourist gaze. *Progress in Human Geography* 21, 359–373.

Davenport, M.A. and Anderson, D.H. (2005) Getting from sense of place to place-based management: An interpretive investigation of place meanings and perceptions of landscape change. *Society & Natural Resources* (18), 625–641.

Eisenhauer, B.W., Krannich, R.S. and Blahna, D.J. (2000) Attachments to special places on public lands: An analysis of activities, reason for attachments, and community connections. *Society and Natural Resources* 13, 421–441.

Farnum, J., Hall, T. and Kruger, L.E. (2005) *Sense of Place in Natural Resource Recreation and Tourism: An Evaluation and Assessment of Research Findings*. United States Department of Agriculture General Technical Report PNW-GTR-660 Forest Service Pacific Northwest Research Station.

Fink, J. (2011) Walking the neighbourhood, seeing the small details of community life: Reflections from a photography walking tour. *Critical Social Policy* 32, 31–50.

Gold, S.J. (2004) Using photography in studies of immigrant communities. *American Behavioral Scientist* 47, 1551–1572.

Haines-Young, R. and Potschin, M. (2009) Methodologies for defining and assessing ecosystem services, Final Report, JNCC, Project Code C08–0170–0062, 69.

Harper, D. (2002) Talking about pictures: A case for photo elicitation. *Visual Studies* 17, 13–26.

Holgate, J., Keles, J. and Kumarappan, L. (2012) Visualizing 'community': An experiment in participatory photography among Kurdish diasporic workers in London. *The Sociological Review* 60, 312–332.

Johnson, S., May J. and Cloke, P. (2008) Imag(in)ing 'homeless places': Using auto-photography to (re)examine the geographies of homelessness. *Area* 40, 194–207.

Kaltenborn, B.P. (1998) Effects of sense of place on responses to environmental impacts. *Applied Geography* 18, 169–189.

Kennard, M. (in prep) *Sense of Place, Tourism and Inshore Fishing in The English Channel and Southern North Sea*. University of Greenwich Unpublished PhD Thesis.

Kerstetter, D. and Bricker, K. (2009) Exploring Fijian's sense of place after exposure to tourism development. *Journal of Sustainable Tourism* 17, 691–708.

Kianicka, S., Buchecker, M., Hunziker, M. and Müller-Böker, U. (2006) Locals' and tourists' sense of place. *Mountain Research and Development* 26, 55–63.

Kyle, G. and Chick, G. (2007) The social construction of a sense of place. *Leisure Sciences* 29, 209–225.

Malpas, J. (2008) New media, cultural heritage and the sense of place: Mapping the conceptual ground. *International Journal of Heritage Studies* 14, 197–209.

Markwell, K.W. (2000) Photo-documentation and analyses as research strategies in human geography. *Australian Geographical Studies* 1, 91–98.

MEA (2005) *Ecosystems and Human Well-being: Synthesis*. Island Press, Washington DC.

Milcu, A.I., Hanspach, J., Abson, D. and Fischer, J. (2013) Cultural ecosystem services: A literature review and prospects for future research. *Ecology and Society* 18 art: 44.

Purcell, R. (2007) Images for change: Community development, community arts and photography. *Community Development Journal* 44, 111–122.

Relph, E. (1976) *Place and Placelessness*. Pion: London.

Rose, G. (1992) Geography as science of observation: The landscape, the gaze and masculinity, in Driver, F. and Rose, G. (eds) *Nature and Science: Essays in the History of Geographical Knowledge*. Historical Geography Research Series, London, pp. 8–18.

Sagoff, M. (2011) The quantification and valuation of ecosystem services. *Ecological Economics* 70, 497–502

Shan, M. and Swinton, S.M. (2011) Valuation of ecosystem services from rural landscapes using agricultural land prices. *Ecological Economics* 70, 1649–1659.

Smith, A. (2014) AHRC looks for ways to welcome artists into the fold. R*esearch Fortnight, 9th July* http://www.absw.org.uk/files/Research_Fortnight_winning_piece_by_Kate_Szell_Dragons_Den_ukcsj14.pdf

Sontag, S. (1977) *On Photography*. Penguin Books: London.

Stedman, R.C. (2003) Is it really just a social construction?: The contribution of the physical environment to sense of place. *Society and Natural Resources* 16, 671–685.

Stewart, W.P., Liebert, D. and Larkin, K.W. (2004) Community identities as visions for landscape change. *Landscape and Urban Planning* 69, 315–334.

Suchar, C.S. (1997) Grounding visual sociology research in shooting scripts. *Qualitative Sociology* 20, 33–55.

Thrift, N. (2008) *Non-Representational Theory: Space, Politics, Affect*. Routledge: London

Tuan, Y.-F. (1974) *Topophilia: A Study of Environmental Perceptions, Attitudes and Values*. Prentice Hall: New Jersey.

UKNEA (2011) *The UK National Ecosystem Assessment: Synthesis of Key Findings*. UNEP-WCMC, Cambridge.

Urquhart, J. and Acott, T. (2013) Constructing 'The Stade': Fishers' and non-fishers' identity and place attachment in Hastings, south-east England. *Marine Policy* 37, 45–54.

Urquhart, J. and Acott, T. (2014) A sense of place in cultural ecosystem services: The case of Cornish fishing communities. *Society and Natural Resources* 27, 3–19.

Van Auken, P.M., Frisvoll, S.J. and Stewart, S.I. (2010) Visualising community: Using participant-driven photo-elicitation for research and application. *Local Environment* 15, 373–388.

Ward, D. (2004) *Landscape Within: Insights and Inspiration for Photographers*. Argentum: London.

Wells, L. (2011) *Land Matters: Landscape Photography, Culture and Identity*. I.B Tauris: London.

Wiley, J. (2007) *Landscape*. Routledge: London.

Williams, D.R. and Stewart, S.I. (1998) Sense of place: An elusive concept that is finding a home in ecosystem management. *Journal of Forestry* 96, 18–23.

Chapter 4
Evaluation, Photography and Intermediation: Connecting Birmingham's Communities

Dave O'Brien

> … Ultimately at times you delude yourself about working collaboratively, you're the one selecting which images they take or whatever, put into a sequence and you're still the main photographer and taking the credit for that. (*Some Cities* project team member)

Introduction

This chapter discusses the evaluation of a participatory photography project, *Some Cities*, which took place in Birmingham, UK, in 2014. *Some Cities* consisted of a series of workshops, public talks, courses on photography along with an online archive and an exhibition at the Library of Birmingham. It was funded from a range of sources, including Arts Council England, University of Birmingham, Birmingham City Council and several local arts organisations. The chapter uses the evaluation to make three points: about participation in art projects; about the relationship between identity and evaluation; and finally about cultural intermediaries.

The chapter raises important points about the practice of evaluation, in particular participatory or co-produced evaluations. The work with *Some Cities* was part funded through a project, *Cultural Intermediation: Connecting Communities in the Creative Urban Economy.* This is one of three major projects on the creative economy funded through the UK Arts and Humanities Research Council's *Connected Communities* research programme. Much of the work within *Connected Communities* has focused on participatory or co-produced research, both in terms of methodological innovation and as an attempt to destabilise the power relationships between community, policy and academic practice. Co-production, as discussed in the following section of this chapter, has two purposes. It is seen as a means to address the inequalities of power in previous community interventions, whether by academia or by government. At the same time the methods advocated by *Connected Communities* stand in contrast to the methods that have been the basis for much of the discourses surrounding community since the 1980s in the UK. Thus the use of participatory or co-produced methods is to do with overturning the intersection between policy and a specific form of social science that underpins and creates government action, as much is it is to do with developing or empowering communities.

The approach described in this chapter stands in contrast to the construction of places, or more specifically areas, as those of deprivation or exclusion. These sorts of constructions are grounded in the social science of the demographic or similar statistical tools. The mode of social science outlined in this chapter reflects both the creative methods recently developed in geography, in particular the role of photography, along with the assumption that project participants are best places to evaluate the impact of an intervention. Most notably the chapter reflects the introduction to the book, whereby the caution against seeing culture as a panacea to urban ills is buttressed by insights into the nature of, and limitations to, forms of evaluation that are not part of the approaches that have become intertwined with advocating participation in culture as an unproblematic route to social transformation. It is in this context that the evaluation of *Some Cities* took place.

The evaluation project recruited 10 participants to be evaluators for *Some Cities*, resulting in five data sets gathered by the evaluators and five follow up interviews by the academic team. Participants were recruited on a voluntary basis and were given an introductory session on evaluation along with some suggested questions, including 'how did you get involved in the project?' and 'what has happened to you as part of the project?'. They were also provided with a voice recorder. They were then free to gather data, ask questions and evaluate in whatever manner they chose, in particular focusing on what they felt was important about *Some Cities*. This data formed the basis of the discussion of *Some Cities* in this chapter.

The chapter confirms the extensive (Carnworth and Brown, 2014) research on participation in arts projects, suggesting that *Some Cities* generated social impacts. These impacts are directly related to the aesthetic element of *Some Cities*, suggesting the specific artistic practice, photography, was crucial to the generation of impacts for participants, in whatever form. This subject is introduced in the chapter's discussion of photography and the evidence of the evaluation described in the section that follows.

The most crucial point made by this chapter is an insight into the relationship between individual identity and the meaning of evaluation. As the discussion in the penultimate section of the chapter shows, there are multiple meanings of evaluation giving differing insights into both the evaluation and *Some Cities*. There is a crucial element of stratification to the modes of evaluation, pointing towards tentative research conclusions about the relationship between forms of class and gender identity and the influence of these structural factors on the meaning and method of evaluation in a participatory context. The chapter has anonymised the data presented, as a result of both the small sample size and the relationship between the data and the individual identity of the participant.

Finally, in the conclusion, the chapter makes a point about the nature of cultural intermediaries. This subject is significant for both recent literature in the sociology of culture (Smith Maguire and Matthews, 2014) and the research project that part-funded *Some Cities*. The evaluation shows a new meaning for cultural intermediation: rather than being about the transmission of elite forms of cultural practice to an emerging middle class and the associated forms of class distinction (Bourdieu, 2010), *Some Cities* offered the possibility that artistic or creative projects may intermediate between citizen and place. The relationship between Birmingham, the project participants and the cultural practice of photography, filtered through the lens of both a mass participatory project and the affordances of digital technology, offers a new way of considering the role of cultural intermediaries that will be important to both the academic literature in this area along with policy and funding decisions.

Evaluation: A Problem for Participation?

Drawn from the study of social policy, Sherry Arnstein's original (1969) ladder of participation remains useful as a starting point for thinking about cultural projects. Her normative typology conceived of participation as a scale, including a range of citizen involvement, from the non-participation of involvement as a type of manipulation through to the expression of citizen power that comes from partnership and citizen control. In the weaker forms of participation citizens are used to legitimate decisions or are included in a tokenistic fashion. This is, of course, not participation at all, but can be badged as such in the interests of the already powerful, such as governments or businesses. The ladder has been adapted across a range of practices, particularly when working with those who are often seen to lack the same capacities as the citizen as originally envisaged by Arnstein, for example Hart's (1997) work with children.

However these initial visions of participation, specifically around political power and control, have been subject to various critiques, suggesting that participation is a form of co-option into the agendas of the already powerful. In the case of cultural or artistic projects, Jancovich (2011) has been especially strident, suggesting that participatory cultural policies, notably those enacted by Arts Council England, are in fact nothing of the sort. The key lesson from Jancovich's work for this chapter is the way in which participatory cultural or artistic programmes may fail to resolve the unequal distribution of power and expertise within projects. At worst this can lead to cultural or artistic outputs that are not only alienating to potential participants, but also may be created in direct and outright hostility to them. The issues Jancovich identifies are particularly acute in those projects that are geographically rooted in seemingly marginalised or excluded communities.

The types of work identified by Jancovich (2011) often leave unresolved the tension, identified in the opening quote, between expertise and participation. Expertise can often be grounded on forms of exclusion and the creation of particular 'domains' (Collins, 2013), whilst participation is often related to co-production that seeks to either remove or at least renegotiate the boundaries created by expertise. Jasanoff (2004) is especially useful at this juncture. Following Latour, she suggests that we view much of social scientific knowledge as already co-produced, grounded both in differing disciplinary forms of knowledge and differing political enterprises buttressed by what is afforded by any given technology. This wider view of co-production can be the starting point for a recognition that the academic expertise within an evaluation is not of a different form or nature to the expertise of the participant. Rather the academic expertise is dependent for its very existence on the participants' knowledge. In this sense what follows, although an academic driven analysis of the nature of evaluation and identity, is a narrative of a research project in which the data was conceptualised and gathered by research participants and the evaluation framework and analysis by the academic team. For these reasons there is a loose co-production of the evaluation as a whole.

This co-production is not along the lines of much of the embedded, participatory action research paradigm that has driven many *Connected Communities* projects. However this looser approach to participation and co-production is useful because it both challenges the understanding of what an evaluation is, as well as drawing attention to the interaction between arts, identity and social scientific research, an important aspect of thinking about co-production on Jasanoff's (2004) analysis. Thus the chapter hopes to add to the growing body of practice of participatory and co-produced research from the perspective of evaluating projects that employ these ideas.

The Nature of Photography

Photography is a research method that can supplement existing ways of doing social science, particularly for marginalised groups (Johnsen et al., 2008, Fink 2012). However what is pertinent here is not the discussion of photography as a research method, which was not the focus of the study, but rather its potential as a participatory medium. This aspect of photography was central to *Some Cities*' mass participatory work as well as to other aspects of the programme. Before considering this, it is worth making a general point about the contested nature of photography, both because this matters to the research findings, but also because it has implications for photography as a cultural practice.

Photography occupies a problematic position vis-à-vis the art world as a whole. It has been especially significant for several continental social theorists, including Barthes, Baudrillard and Bourdieu. These theorists, along with a variety of others, were important in establishing photography as a legitimate art form with associated theorists and theorisations. However photography also exists as a daily, everyday practice that often has little to do with conceptions of elite art, but rather is bound up with an aesthetic that is more to do with everyday life.

This tension plays out notably in the work of French sociologist Pierre Bourdieu, an academic whose work is influential on the understanding of British cultural consumption and participation (Bennett et al., 2009). Alongside Bourdieu's methodological innovations, his work allowed British sociology to more clearly articulate the relationship between culture and forms of social division. These forms of social divisions, based on cultural practices, were found in photography as much as they were found in museum attendance or classical music listening. For Bourdieu (1996) the mass embrace of photography was made possible by the types of technological development that were synonymous with Kodak film processing. These developments would, of course, come to be superseded by the types of devices, specifically smartphones, which were essential to *Some Cities* mass participatory elements. However, photography, for Bourdieu, was not the mass medium available to all that it seemed. Rather its use was socially stratified (in the France of the 1960s that was the subject of his analysis), with specific attachments and divisions around artefacts such as the family portrait or the artistic photograph. Fundamentally photography occupied a problematic social status, not fully embraced by the elite of the art world, but obviously stratified around levels of aesthetic composition and appreciation (Stallabrass, 1996).

The technological transformation of photography has driven this stratification, particularly as the smartphone has led to an explosion of digital photography that is publically exhibited online using spaces such as Instagram. Both Stallabrass (2013) and Gye (2007) have drawn attention to the transformations associated with the camera phone. For Stallabrass the boundary making associated with expertise referenced in the previous section has been important for 'art-world' photographers who defend their technical skills. This is in the context of the rise of digital technologies that mean: 'It is not only harder to say whether or not you are a photographer, but harder to say where the lines lie between professional, social or artistic work' (Stallabrass, 2013: 4).

The blurring of lines around the work of the photograph is paralleled in discussions of the social practices attached to photography identified by Bourdieu (1996) and examined by Gye (2007). Gye (2007: 285) cites Okabe and Ito to identify the impact of the camera phone on the social function of photography, whereby:

> The social function of the camera phone differs from the social function of the camera in some important ways. In comparison to the traditional camera, most of the images taken by camera phone are short-lived and ephemeral. The camera phone is a more ubiquitous and lightweight presence, and is used for more personal, less objectified viewpoint and sharing among intimates. Traditionally, the camera would get trotted out for special excursions and events – noteworthy moments bracketed off from the mundane ... The camera phone tends to be used more frequently as a kind of archive of a personal trajectory or viewpoint on the world, a collection of fragments of everyday life.

The collection of everyday life, by an ever-present technology, is crucial to how new media reshapes understandings of place. In this sense photography is best viewed less as a stratified cultural pursuit (although that is ever present for understanding culture) but rather as a pervasive media that has implications for the construction of sense of place within the built and lived environment, and thus for individual and community identity:

> New media can assist in returning us to a sense of place that involves more than simple location or displaceable 'information' (this may also be where game engines have an 'analytical' or 'experimental' role to play). It can enable us better to understand the ways in which we do indeed belong to the places in which we live, and have lived, and the ways in which a sense of culture and identity, and with it a sense of heritage, must also entail a sense of place. (Malpas 2008: 207)

Understanding *Some Cities*

This final point, on the implications of photography as a pervasive and social media, was crucial to the mass community involvement elements of the *Some Cities* project, in particular the use of twitter hashtags #brumphotos and #somecities. In some respects this makes *Some Cities* easy to characterise and evaluate. The project received positive local media coverage from local press and television, along with substantive online participation. For example by March 2014 the online archive had collected over 40,000 photographs. This main website was linked to a Tumblr archive consisting of 2556 unique entries by October 2014. Once the Tumblr was superseded by www.some-cities.org.uk the collection grew fuelled by the site allowing for much easier uploading of images, primarily via the twitter hash tags.

In other respects the evaluation is more complicated. The online elements raised the issue of aesthetic quality as opposed to narrative or expressive photography of the sort discussed in the previous section's consideration of both the stratification of photography and the impact of digital camera phones. Indeed the breadth of the project, combining the elements of community and personal development with the mass participation and formal exhibition programmes, makes a comprehensive evaluation almost impossible given the different methods needed to evaluate each element. Moreover, given the stress on co-production and participation within the *Connected Communities* programme it is worth considering the criteria for assessing the scope of *Some Cities* offered by one of the interviewees from the project. *Some Cities* was thus as much to do with the 'unforseeable and imagined activities' as it was to do with the formal programme proposed to the funders.

As a result the reminder of this chapter concentrates on the findings from the participatory evaluation of individuals involved in the photography classes. The first finding focuses on *Some Cities* as a form of community art. The relationship between community art and photography is summarised in Purcell's (2009: 115) work. Here photography in community art settings can take three forms:

> As formal or non-formal courses delivered in a community setting mainly as skill development for individuals, through worker-led initiatives such as 'photographer in residence' schemes that may or may not have a development purpose, or as a community driven initiative with a defined development objective.

Some Cities was not directly aimed at community development, but did take up the most common model, that of skill development, as part of its programme of work. As discussed below *Some Cities* created many of the positive impacts associated with projects of this type, including skill development, increased social capital and a sense of increased civic pride (Purcell, 2009; Carey and Sutton, 2004).

One starting point for thinking about the experience of participants in *Some Cities'* workshops is their comments when interviewed following the end of the evaluation. They were strongly positive about the activity, regardless of their levels of expertise or experience. Three quotes illustrate this point:

> I find it really enjoyable, the whole journey, the lectures that we had were quite cool as I learnt some new stuff, even though I did an A-Level in photography.

> It's been really fun.

> I'm really happy with everything that its taught me so far, I'm over the moon with the media response, the way people across Birmingham have taken to the project have got behind it and understood it.

This last quote broadens out the conception of success for the project beyond the specific activities of the classes and lectures. The aim of *Some Cities* to involve the population of Birmingham in a mass-participation photography project was perceived as highly successful by this respondent. Indeed the relationship between participants and Birmingham was crucial to the participants' ideas of success, as increased civic pride was expressed:

> I'm proud of what they (*Some Cities*) are doing. They are making Birmingham start to stand out, we need that, we get a lot of stick us Brummies do, especially for our accent.

> So many things I've missed in the city I've lived in all my life and even now keep noticing new things, thinking, oh god, I've never seen that before.

> The project has civic pride running through its blood.

Along with further comments on the sense of mass participation identified in the original set of quotes:

> I found *Some Cities* was all of a sudden lots of people being interested in Birmingham and I wanted to be part of it.

Here is evidence of the type of effect Malpas (2008) identified as sense of place and space is intertwined with what the mass medium of various digital devices affords.

The social elements of the participants' perception of the online programme were also reflected in the individual social impact of participating in *Some Cities*.

There is a wealth of literature suggesting that engagement with artistic or cultural projects can have positive social impacts (Matarasso, 1997). Indeed this ranges from health benefits through to impacts on community cohesion and regeneration (Evans and Shaw, 2004). For the participant evaluators the elements of civic identity discussed previously were extended by developing a sense of social capital and connectedness and the social elements associated with attending class and being part of the shared process:

> It's been great to meet everybody … . It's the social aspect of it … it seems a shame as we've just come to the end and you're comfortable now and you want to carry on.

The activity of learning a technical skill (reflecting the debates outlined previously) was also important to building participants' confidence, a theme that was common across the five respondents, reflected in comments such as:

> The sharing, the understanding of the basics, the being confident, that all came with the course and I tried to apply that, that came through the sharing, it came through the listening and learning, and I tried to apply that to my own project … . and I still do.

And further examples of social capital development:

> If it wasn't for my camera, and getting to know the community and getting involved in the community here as a committee member [for a local cultural event] I probably would have had another celebration where I'd stayed in my flat and kept away from it all. It's sort of given me confidence getting involved in the community and getting my camera to get out and about.

The experience of enhanced confidence is common to much arts participation (Carnworth and Brown, 2014). In the case of *Some Cities* this was grounded in the technical aspects of photographic practice, whereby acquiring technical mastery of the cameras, the developing process and the aesthetic elements of photography was central to the expression of participants' confidence:

> I've learned to trust my camera more, I've given it a name … and I feel more confident going out, composing shots and understanding the basics of digital photography.

For Stallabrass (2013), the influential writer on photography, it is now problematic to claim the status of the photographer in the face of the further democratising of the media by the digital revolution. The discussion of photography by the participants reflects this problematic, both in terms of the acquisition of expertise, but also in terms of the assumptions around the accessibility of photography and its status as a mass cultural practice:

> It's a humbling experience and quite an eye opener … [it] really made me realise there's a lot of technical ability, probably sounds a bit arrogant, but I thought photography was quite easy!

This is also related to the individual narratives of identity in the evaluation data, whereby the sense of self, of expertise and specific personal experiences were drawn upon to reflect the impact of *Some Cities*. This example brings together these themes:

> I tried to haggle, a flea market in Paris … . and it sounded alright but when I asked him how much he wanted, he wanted 80 euros, and I just said to him no and I don't think he realised that I knew what I was trying to look for.

The sense of confidence and expertise was also connected with the participants' relationship to Birmingham, which became a site for the expression of confidence and social connectedness. Gye (2007) has raised important points about the use photography as a social medium. They are reflected in the participants' discussion of Birmingham, where personal relationships are related to understandings of how the social and identity is constructed, developed and displayed by the act of taking, selecting and displaying the photograph(s). This was seen in all of the participants' data, reflecting the personal nature of their assignments for class, as well as their sense of self. Their photography projects included journeys to work or locally based explorations:

> Showing how everyone's united in the community … I thought that was quite important to show how different cultures and diversities of people are all coming together.

As well as projects concerned with the aesthetics of the image produced. The influence of personal identity was mirrored in the participants' evaluations.

Evaluation and Identity

The construction, development and display of identity, in all its complexity (Lawler, 2014) was fundamental to the most interesting finding of the evaluation. This was not specifically concerned with *Some Cities'* impact, but rather it was concerned with the nature of evaluation in the context of limited participant control over the collection of data.

The evaluation suggested much of the expected outcomes for an arts project (Carnworth and Brown, 2014). Indeed, the evaluation method reinforces the range of positive outcomes because of the investment of time and the commitment to data collection exhibited by the participants. The nature of the sample (N=5) is vital here. It would be inappropriate to make policy or future funding decisions based on the assessment of the success or failure of *Some Cities* based on the data presented above. What the data can do is point to the usefulness of participatory evaluation methods and the benefits of recognising the co-produced nature of this evaluation. When this perspective is considered the sample, albeit small, offers a considerable insight into the relationship between the understanding of evaluation and the nature of participants identities. Indeed it supports the previous sections' attentiveness to both the identity work necessary to feel successful within the artistic medium of photography and begins to account for the appreciation voiced for *Some Cities*' social and civic impact.

There were three versions of evaluation, with three distinct meanings and three differing forms of data presented by the participant evaluators. These were clustered around: personal narratives, with diary keeping and self-interviews; a research project, with more formal interviews and research questions; and a record, with recordings of the participants' classes.

In the case of the personal narrative, evaluation was personalised to the impact on the evaluator, expressed in the form of self-interviews and diary entries. For this mode of evaluation the attention of the evaluator was focused on their relationship to the other participants, with the questions of social connectedness and confidence building paramount:

> It created a bit of a bond with me and one of the other students who was going through the same process, so afterwards we went for a drink to celebrate … . there's a lot to it.

Evaluation here is therefore an expression of introspective personal identity and reflects the identity work discussed in the previous section. This is related, but obviously differs to the other two understandings of evaluation expressed as part of the research.

For another participant the act of interrogation was the essence of evaluation, manifested by interviewing and questioning the project leaders and other participants of *Some Cities*. Questions were common such as:

> Are you saying that there's some trust that's been built up with the *Some Cities* project … some trust and actually you are inspiring people in the way that you wanted to inspire them?

In this version of evaluation the internalisation of social scientific concepts and approaches, described by Savage (2010) as a crucial component of British, specifically middle class British, identity, is clear. Here evaluation is not means to judge the impact of participation in *Some Cities* upon oneself by explicitly offering a personal narrative. Rather the evaluator in this understanding is an externally focused researcher, gathering data with which to make judgements about the success or failure of the artistic programme of which they are a part.

The final manifestation of the meaning of evaluation was as a factual record of the events of the classes. This approach was most distant from the introspective judgements of impact on the person and rather sought to present data as 'the facts' of what had occurred in *Some Cities*. Within this version of evaluation it is clear that attendance and learning are the most important aspects of *Some Cities*, but grounded in a record of participants' apprehension as to their success or failure in mastering the technical activity of photography: 'That's the only problem with learning is the cost, especially when there's waste'.

Moreover this approach to evaluation revealed the importance of the mundane and the everyday in establishing the social capital and connectedness discussed in the previous section, whereby the struggle for mastery of the techniques of developing film were accompanied by the most central of British social rituals: 'I think we need to put the kettle back on really'

These differing interpretations of evaluation raises several issues for both the project of which the work with *Some Cities* was a part as well as future uses of participatory evaluation on other projects. Obviously it is essential to be exceptionally cautious when thinking sociologically from such a small sample size. However (and notwithstanding the levels of anonymity granted to the participants within this chapter) there were signals of the stratification of evaluation, both around educational background, which is of course closely related to class and social status, and gender.

Without compromising the anonymity of the participants, it is possible to observe how the introspective activities of self-interviews and diary keeping where closely associated with the female participants in the evaluation. This reflects Bennett et al.'s (2009) work on the stratification of culture in the UK, whereby gender is an important factor in structuring the consumption of inward facing or introspective activity. The male participants reflected both the forms of social scientific identity described by Savage (2010) and a concern with 'the facts' of the project, as opposed to telling of an individual story.

The tentative suggestion of such a pattern raises a potential direction for future research seeking to employ participatory evaluations, particularly those that use qualitative methods. Over and above any assessment of the projects under evaluation there is the opportunity to learn about evaluation itself. The data gathered by the evaluators, organised and analysed by the academics and presented within this chapter suggests a plurality of understandings and meanings of evaluation itself. Thus we return to Jasanoff's conception of the social world, and the expertise within it, as co-produced, but we may ask the question as to the stratification of that co-production in the context of participatory evaluation.

Conclusion

The note of caution sounded in the previous section is worth continuing as a form of conclusion. To an extent there is maybe little to be said from a small sample, looking at only part of the project. Indeed, from the point of view of policy this approach to evaluation looks problematic (Nathan 2015, in press), with little evidence of the positivistic economics-based frameworks that are demanded in public policy (O'Brien, 2010). However this experimental type of evaluation has benefits that are of a different kind to those associated with the randomised control trial or the cost benefit analysis. As this chapter has shown, participatory evaluation gives access to the subjective experiences of the participants, as well as containing lessons about the relationship between identity and the practice of evaluation itself. The act of co-production of the evaluation, even in its loosest form, can thus show us how the social world is constructed (Jasanoff, 2004). It is this idea that allows the research to make a more general point about cultural intermediaries, a topic that was crucial to the research project funding part of *Some Cities*. Indeed this is present in the quote that opened this chapter, whereby the Bourdieuian reading of the expertise of the cultural intermediary is still present, whilst the project itself reveals a very different role and form.

The cultural intermediaries, in this case the *Some Cities* project team, offer a controlled challenge to the aesthetic hierarchies that have come under sustained assault as a result of the emergence of photography as a mass, digital, cultural form (Stallabrass, 2013; Gye, 2007). Based on the discussion in this chapter it is possible to argue that the moment of intermediation provided by *Some Cities* is not between legitimate and illegitimate forms of culture, but rather between the population of Birmingham and its city. This comes both from the evidence of the evaluators, but also through the evidence from the web project, whereby the everyday, often mundane, cultural practices of the city, such as the night out, the journey to work or the family photo, are placed in the context of an art project. The social divisions identified by Bourdieu (1996) on photography and cultural intermediaries have not disappeared. Rather, the structures of the art world (Becker, 2008) and the structures of the everyday are brought closer together in *Some Cities*. This act of intermediation is one that, when placed in the context of the stratified nature of cultural funding in the UK, can be seen as having true social importance.

Acknowledgements

This chapter emerges from research undertaken as part of the *Cultural intermediation: connecting communities in the creative urban economy* project funded by the Arts and Humanities Research Council (grant reference AH/J005320/1). More information about the project can be found at culturalintermediation.org.uk.

References

Arnstein, S. (1969) A ladder of citizen participation. *Journal of American Institute of Planners* 35(4), 216–224.

Becker, H. (2008) *Art Worlds*. Oakland: University of California Press.

Bennett, T., Savage, M., Silva, E., Warde, A., Gayo-Cal, M. and Wright, D. (2009) *Culture, Class and Distinction*. London: Routledge.

Bourdieu, P. (1996) *Photography: A Middlebrow Art*. London: Polity Press.

Bourdieu, P. (2010) *Distinction*. London: Routledge.

Carey, P. and Sutton, S. (2004) Community development through participatory arts. *Community Development Journal* 39(2), 123–134.

Carnworth, J. and Brown, A. (2014) *Understanding the value and impact of cultural experiences* available from http://www.artscouncil.org.uk/media/uploads/pdf/Understanding_the_value_and_impacts_of_cultural_experiences.pdf accessed 21/8/2014.

Collins, H. (2013) Three dimensions of expertise. *Phenomenology and the Cognitive Sciences* 12(2), 253–273.

Evans, G. and Shaw, P. (2004) *The Contribution of Culture to Urban Regeneration in the UK*. London: DCMS.

Fink, J. (2012) Walking the neighbourhood, seeing the small details of community life. *Critical Social Policy* 32(1), 31–50.

Gye, L. (2007) Picture this: The impact of mobile camera phones on personal photographic practices. *Continuum: Journal of Media and Cultural Studies* 21(2), 279–288.

Hart, R.A. (1997) *Children's Participation: The Theory and Practice of Involving Young Citizens in Community Development and Environmental Care*. London: Routledge.

Jancovich, L. (2011) Great art for everyone? Engagement and participation policy in the arts. *Cultural Trends* 20(3–4), 271–279.

Jasanoff, S. (2004) *States of Knowledge: Co-production of Science and the Social Order*. London: Routledge.

Johnsen, S., May, J. and Cloke, P. (2008) Imag(in)ing 'homeless places' *Area* 40(2), 194–207.

Lawler, S. (2014) *Identity*. London: Polity Press.

Malpas, J. (2008) New media, cultural heritage and the sense of place: Mapping the conceptual ground. *International Journal of Heritage Studies* 14(3), 197–209.

Matarasso, F. (1997) *Use or ornament* London: CoMedia.

Nathan, M. (2015 in press) What works in urban regeneration? in Matthews, P. and O'Brien, D. (eds) *After regeneration*. Bristol: Policy Press.

O'Brien, D. (2010) *Measuring the Value of Culture*. London: DCMS.

Purcell, R. (2009) Images for change: Community development, community arts and photography. *Community Development Journal* 44(1), 111–122.

Savage, M. (2010) *Identities and Social Change in Britain since 1940: The Politics of Method*. Oxford: Oxford University Press.

Smith Maguire, J. and Matthews, J. (2014) *The Cultural Intermediaries Reader*. London: Sage.

Stallabrass, J. (1996) 'Cold eye'. *New Left Review* 220, 147–52.

Stallabrass, J. (2013) *Are you a photographer?* http://photoworks.org.uk/are-you-a-photographer/ accessed 6/11/14.

Chapter 5
Creative Place-making: Where Legal Geography Meets Legal Consciousness

Antonia Layard and Jane Milling

Introduction

This chapter investigates how creative participation in place-making is legally constructed. It draws on the findings of the AHRC *Creative Participation* project, which explored how three 'pioneer communities' (the Elders Council of Newcastle, Northern Youth[1] and the People's Republic of Stokes Croft (PRSC) in Bristol) use creativity to involve themselves in place-making and planning practices. Each of these groups is working to improve their locality, albeit in quite different ways. While all began by working through formal consultative and participatory procedures, each found that voices are 'not heard', so that you 'have to use every avenue that you can'.

The research asks why some participants engage with developers through formal planning processes or paint graffiti without permission, while others feel unable to act. It suggests that one reason for the differences, is that in addition to distinct sets of resources, there are variations in 'legal consciousness' (Ewick and Silbey, 1998), differences in the ways in which participants' social and cultural practices enact legality. To use one leading formulation, consciousness is understood here as 'part of a reciprocal process in which the meanings given by individuals to their world become patterned, stabilized and objectified. These meanings, once institutionalized, become part of the material and discursive systems that limit and constrain future meaning making' (Ewick and Silbey 1998: 39). Legal consciousness, then, considers how law (very broadly conceived) and understandings of legality or illegality, shapes, frames and categorises social life (Ewick and Silbey, 1998). To understand creative place-making, and the ways in which the legal, the social and spatial interact, this chapter suggests that we need to investigate how participants understand legal processes and practices (their legal consciousness) in order to explore how these divergent framings of legality lead to such diverse approaches to place-making.

Bringing in legal geography lets us work from the site up, drawing in both 'law in books' as well as co-constitutive practices of legality in society (Blomley 2004; Delaney, 2010; Braverman et al., 2014). Places are relational and they are legally

1 Some names have been changed.

made and re-made both as large, one-off construction projects (through property, planning, compulsory purchase law, and so on) as well as on a daily basis (through rules of property, criminal or insurance law, to name but a few). For the purposes of this study, place-making is understood as the planning, design and management of public spaces. These include highways, squares, shopping centres and public amenities, where participants have attempted to change the look, feel or sight of a locality.

In legal geography, places are assumed to be made, not given. The working assumption is that in individual, place-specific locations, the social, the spatial and the legal are co-constituted (Blomley, 2003) and that these are active, 'world-making' processes (Delaney, 2010). Legal provisions and frameworks facilitate relationships between landowners, regulators, police forces as well as those who lack site specific status: not citizens or residents but visitors.

At the heart of place-making, then, is this tension between owners (all places are somebody or some group's property) and 'the public'. There is a longstanding research question in the academy, which asks who 'the public' are in instances of place-making, particularly in large regeneration projects (Cochrane, 1986). Motives for involvement by members of the public can be diverse. Certainly voluntary labour is increasingly explicitly enrolled (Warren, 2014), including in the context of the much heralded (but increasingly forgotten) 'Big Society'. Often however (as here) contributions are initially unsolicited. The variety of these contributions – formal and informal, representative or apparently not – raise the larger question of whether plurality can ever be incorporated into decisions about individual pieces of land (Pollock and Sharp, 2012; Aitken, 2012). How to integrate different voices continues to matter where property law gives the owners the 'agenda-setting' power (Katz, 2008). Understanding legal consciousness can explain how 'publics' engage with landowners and authorities in very different ways. This chapter considers what this can mean for creative place-making.

Legal Geography Meets Legal consciousness

i) Legal Geography

The starting point for legal geography has been a concern that legal practice does not sufficiently allow for the specificity of place. Critics of conventional legal practice have suggested that laws should not be (spatially) equal. They suggest that law should be 'dependent on' rather than 'transcend' place (Bartel et al. 2013: 349). Prompted by this research question, analysts have investigated the reflexivity and the co-constitution of space, law and society. Working from the site up (so that, for example, place, rather than legal rules, can become the focus of research) legal geographers have come up with a series of concepts to capture the interrelationship of the legal, the social and the spatial on the ground.

The foundational conceptual device in legal geography is Blomley's concept of a 'splice' (Blomley 1994, 2003), which shares similarities with Delaney's (2010) 'nomosphere' and with 'lawscapes' (Philippopoulos-Mihalopoulos, 2007; Graham 2010). Splices (like nomospheres and lawscapes) identify instances or moments where legally informed decisions and actions *take place* in the sense both of the occurrence of a legal performative *and* of being spatially located and embodied (Bennett and Layard, forthcoming). They are encodings, which combine spatial and legal meanings (Blomley 2003; Matthey et al. 2013; Layard 2014). Streets, squares, gardens, airports, forests, protest camps, mosques or military command centres can all be understood as splices.

If splices, nomospheres, and lawscapes are the core concepts for legal geographers, how do we investigate them? How can we examine the co-constitution of places and place-making legally, spatially and temporally, as well as discursively and materially? One way, this chapter suggests, is to draw on research on legal consciousness, to understand how understandings of legality can affect why some groups become 'pioneers', engaging with policymakers, property owners and local authorities in both legal and illegal ways. It also explains why other groups of participants co-constitute places more passively, by doing less, for example, enabling the property owner or other actors to have a greater say over the site. The central point is that places are co-constituted: socially, spatially and legally.

One question legal geography has repeatedly asked, is how places are 'made' (Mitchell, 2003; Layard, 2010; Blomley, 2007; Matthey et al. 2013). Analysing place-making through a legal geographic lens, we can say that the property owner governs the space through limiting access (no protestors, no beggars), controlling what can happen where (no skateboarding, no busking), producing the aesthetic of the site (the paint colour, the planting) and governing security (installing CCTV or calling the police for assistance). In addition, however, others, whether legally welcome or not, continually make and re-make places, particularly if they are open to all (or apparently all). Combining legal geography and working from the site up with legal consciousness enables us to investigate how not just law (legal provisions or practice) but also understandings of legality and differences in legal consciousness, contribute to place-making. It lets us consider, how and whether, those distinctions in legal consciousness can or should be harnessed for explicitly creative place-making.

ii) Legal Consciousness

Legal consciousness studies really began with critical legal scholars (CLS) who investigated perceptions of law's dominance and reification through studies of legal decisions and judicial adjudication (Klare, 1978; Kennedy, 1980). Through their work they developed a 'theoretical postulate' of 'law's hegemonic role in sustaining domination' (Halliday and Morgan, 2013: 3). Central to these investigations was the notion of the 'indeterminacy thesis' ('that legal questions lack single right answers', Kress, 1989: 283) and the reification of law. They investigated how, in addition to institutional establishers of hegemony such as the Church, schools and media outlets, legal practice can 'induce submission to a dominant worldview', focusing on an overarching notion of (state) hegemony in the singular (Litowitz, 2000: 2).

Building on this work, and work on legal culture (Silbey, 2010; Cotterell, 1997) as well as insights from Durkheim and Weber, law in society scholars moved sideways, using empirical research to investigate legal consciousness. Theoretically, they also began at a different place from CLS scholars, drawing on both deconstructionist insights and experiences of 'law in action' that they saw in the everyday. As Silbey (2005: 330) explains, law in society scholars assumed that: 'the law is less pliable and less amenable to reinterpretation and reconstruction than poststructural critiques of determinacy seemed to suggest. Indeterminacy does not make all things possible; it means only that possibilities are not predetermined or fixed'. They shared an assumption of hegemony then, but felt it was not as institutionalist as CLS scholars suggested.

Specifically, law in society researchers of legal consciousness were explicit in focusing on how 'legality operates ... as both an interpretive framework and a set of resources with which and through which the social world (including that part known as law) is constituted' (Ewick and Silbey, 1998: 23). Assuming reflexivity between legality and society, these law and society scholars also adopted an assumption of co-constitution, similar to those underpinning legal geography (though this clearly also incorporates the spatial). Rather than adopting a structuralist perspective, this strand of legal consciousness research attempted to bridge structure and subjectivity. In particular, it investigated legal hegemony, asking why people accept unequal and uneven legal practices. It aimed to understand the question of how unequal legal experiences have been made to seem normal and natural? In particular, as law in society scholars keep asking, why do 'the haves' continue to get ahead (Galanter 1974, Kritzer and Silbey, 2003)?

To answer these questions, Ewick and Silbey published a seminal book, *The Common Place of Law* in 1998. Its opening told the story of Millie, a poorly paid black domestic housekeeper. Millie was charged with driving a vehicle uninsured and leaving the scene of an accident that was caused while a friend of her son was driving it without permission. The book starts by comparing Millie's experience of legal practices, including the loss of her driving licence, with her employers' experience of these legal practices when they intervene to help her, challenging the 'offence' and, ultimately, getting the situation resolved.

In their analysis, Ewick and Silbey identified three different types of consciousness: 'before the law' (conformity); 'with the law' (contestation); and 'against the law' (resistance). They documented how, while Millie conformed, believing in the process and being submissive and deferent to the law (Ibid.: 9), her employers contested. For them 'the legal engagement is enacted with a sense of entitlement and routine, one of a variety of experiences used to establish social position, defend personhood, and construct identity' (Ibid.:11). For, when

Millie was first convicted, she resisted. Required by the court to do community service, she arranged for her service to be in the Church where she already did volunteer work. In this way she: 'successfully insinuated her life into the space of the law and, in doing so, reversed for a moment the trajectory of power … with her ruse, she succeeded, where earlier she had failed, to infiltrate the dominant text' (Ibid.: 12).

At the heart of Ewick and Silbey's argument lies a tripartite analysis of the way in which non-lawyers (assuming there is such a thing) engage with legality. The first, *before the law* encompasses understandings where law is an abstract entity removed from everyday life. Here subjects conform:

> In this form of consciousness, the law is described as a formally ordered, rational, and hierarchical system of known rules and procedures. Respondents conceive of legality as something relatively fixed and impervious to individual action. (Ibid.: 47)

This ideal type represents something of Millie's legal consciousness in this setting. As Harding (2006) notes, this form of consciousness, conceptualises 'the law' as a unitary body where 'by going to 'law' to seek a solution to a problem, an individual loses control over the outcome of the situation, as they transfer the 'power' to determine the appropriate solution to their problem to 'the law'. Here the 'usage of the term 'law' operates as a claim to power in that it embodies a claim to a superior and unified field of knowledge' (Harding, 2006, 513 citing Smart: 4). The perception of powerlessness is therefore compounded by a reification of law's ability to 'know' what the correct solution should be. We can see an overlap here with CLS critiques of the determinacy thesis, and of reified understandings of 'the law'.

The second ideal type is *with the law* where the legal system is 'an arena of contest', a game to be played:

> The law 'is an arena of competitive tactical maneuvering where the pursuit of self-interest is expected and the skillful and resourceful can make strategic gains … respondents display less concern about the legitimacy of legal procedures than about their effectiveness for achieving desires'. (Ewick and Silbey, 1998: 48)

This ideal type of contestation is enacted by Millie's employers, when they intervene in the legal process to ask their company attorney to have the case reopened, which it was, after which the court found Millie not guilty and dismissed the charges. The amount of the fine that Millie had already paid, was repaid to her, and her licence was reinstated. Millie's employers had engaged with the law in game-like agonistic context.

The third ideal type is *against the law* where people reveal their sense of being 'up against the law', where law is perceived as a commodity of power, subjective in its application and broadly useless. Here people engage in resistance with a consciousness both of being less powerful and a consciousness of opportunity, 'a situation in which one might intervene and turn to one's advantage' (Ibid.: 183). Resistance is a claim to redress, attributing responsibility by 'making do', fashioning:

> Solutions they would not be able to achieve within conventionally recognized schemas and resources. People exploit the interstices of conventional social practices to forge moments of respite from the power of law. Foot-dragging, omissions, ploys, small deceits, humor, and making scenes are typical forms of resistance for those up against the law. (Ibid.: 48)

In Millie's case, by suggesting her church service as fulfilling her legal requirement to perform community service, she resisted the law, reversing for a moment 'the trajectory of power' (Ewick and Silbey, 1998: 12).

Of course these ideal types are heuristics, not static categories. Ewick and Silbey have noted from the outset that legal consciousness is not 'fixed nor necessarily consistent; rather it is plural and variable across contexts, and it often expresses and contains contradiction' (Ewick and Silbey, 1998: 50). These types of legal consciousness are not mutually exclusive. Consciousness entails both 'thinking and doing': the complexities and contradictions within legal consciousness are what make 'legality' a type of social practice and therefore a producer of social structures (Ibid.: 224–5). This is a consistent insight from legal consciousness scholars who have found (whether amongst working class participants in New Jersey, welfare benefit recipients or clients in divorce offices) 'that legal consciousness is, like law itself, polyvocal contingent and variable' (Sarat 1990: 374 and 375 respectively), acknowledging nevertheless that '[s]uch polyvocality, contingency and variability do not, however, dislodge the power of law or the dominance of legal rules and practices' (Ibid.: 376).

The aim of this law and society legal consciousness research has been to investigate legality and how it contributes to political, social and economic hegemony. The template has been much used to investigate the power of legal effects and the ways in which law and legality reproduce existing power hierarchies. But what of people who feel empowered to effect change outside of these legal processes, moving beyond either contestation or resistance? And what of people acting collectively, rather than as individuals?

Extending this single study of legality then, scholars have decentered 'law' in their analyses. They have studied legality both in privileged sites such as courts, small claims courts (Conley and O'Barr, 1990, Merry, 1990) or lawyers' offices (Sarat and Felstiner, 1986) as well as in 'the everyday' (Ewick and Silbey 1998, Nielsen 2000, Levine and Mellema, 2001). These latter investigations frequently draw on Foucault and de Certeau's work, suggesting that 'investigations of law's power are most fruitful not at the level of legal institutions and the state but at the level of lived experience, where we can see how power is exercised, understood and sometimes, resisted' (Mezey 2001, 145).

Moving the debate forward still further, Fritsvold (2009) in an important paper, addressed these questions by adding a fourth category, *under the law*, to Ewick and Silbey's tripartite distinctions. He used this characterisation to describe radical environmental activists who were not prepared to accept the *status quo*, arguing that these actors perceived 'the law as fundamentally illegitimate because it is created by and embedded in a social order that is fundamentally illegitimate' so that this category of legal consciousness:

> Perceives that a vitally corrupt social order has produced a vitally corrupt legal system to assist in perpetuating its own power … . Against the Law observes that the law often fails as an asset to achieve justice; Under the Law views this failing as intentional and perceives law as an active agent of injustice … radical environmental activists often engage in flamboyant acts of instrumental lawbreaking for the purpose of symbolic or actual subversion. They purposefully, and often very visibly, break the law and openly challenge the legitimacy of law and the social order – hence Under the Law. (Fritsvold, 2009: 810)

While Fritsvold's category is presented as a fourth dimension of legal consciousness work, it actually takes a different tack. Rather than considering how legality contributes to hegemony, Fritsvold investigates *il*legality and how this type of legal consciousness contributes to *challenging* hegemony. His focus is on the activists, trying to understand how and why collective voices of dissent are distinguishable from other 'everyday' actors. His subjects are people who do not accept the hegemony, yet rather than engage 'with the law', as Millie's employers did, they move out of the (conventional) everyday and protest, acting differently, effectively 'out of place' (Cresswell, 1996). Fritsvold's focus then is on *il*legal consciousness and *il*legality.

Participants with an 'under the law' consciousness are not conforming to, contesting or resisting 'the law', they are *repositioning* themselves, conforming with a 'higher law'. They are 'making the court a kind of pulpit for preaching a vision of an alternative world, even while it continues to carry out the work of resolving conflict in the world as it is presently' (Halliday and Morgan, 2013: 26), viewing it 'as a crucial place for articulating an alternative moral, political, and social imaginary' (Ibid.: 25). Illustrating the dynamic nature of legal consciousness, there is often an acceptance of some, if not all, legal provisions and practices, for example, the right to protest. Investigating interviews with radical environmental activists in the UK, Halliday and Morgan found: 'implicit recognition of the normative value inhering in a formal legality that is – if only sometimes and then unpredictably – holding to account the exercise of public power by public officials' (2013: 24).

As this chapter illustrates, our participants also demonstrated different legal consciousnesses in relation to place-making. The two groups that have been most successful are the Elders in Newcastle who 'contested' and were 'with the law', and the creative activists in Bristol who repositioned themselves, urging beautification 'under the law'. Our younger participants belonging to Northern Youth, who were much less well resourced, while they also repositioned themselves creatively, particularly on their trip to Brussels, generally 'conformed'. They found themselves largely unable to place-make outside of their boundaries of the properties owned by the youth [service]. All three groups, however, realized the limitations and used innovation, humour and resourcefulness to step outside legal norms. All at some times were 'under the law', working to 'higher' goals. What the case studies tell us, is that understanding this different form of legal consciousness, 'under the law', can provide seed corn to promote creative place-making more widely.

Creative Participation: Three Case Studies

The project Creative Participation was funded by the AHRC under its Connected Communities project. It enabled the authors to collaborate with representatives of three communities, the Elders Council of Newcastle, Northern Youth, and the People's Republic of Stokes Croft (PRSC) in Bristol, to investigate how creative place-making occurs in different sites and with different age groups. The project consisted of interviews, observations, walking tours and a workshop drawing together contributions from each of the groups. An exhibition, *Creative Citizen*, curated by PRSC, was also held in Stokes Croft from January – February 2012, which was the focal point for the workshop and attracted many visitors from Bristol.

The first of the three participant groups, the Elders Council of Newcastle, is the older people's Council run by and for older people in Newcastle. In addition to projects on quality of life, loneliness and social inclusion, the Council has an Older Person Friendly City Group, which has a programme of work on assessing older person friendly environments. Using a checklist based on members' views on what is important to older people, the working group has carried out systematic assessments of Newcastle City Centre, Parks and Recreation Areas, Supermarkets and the District Centres in Newcastle. Systematic reports are produced for each inquiry including sections on 'method', 'recommendations' and 'summary of findings' and, in some, copies of the questionnaire used. Each report is produced in large font to be easily readable by older people, while some are illustrated with very funny cartoons. This work has been supplemented by creative involvement as well, including by the 'Old Spice' theatre group, begun as an offshoot of the Health Action Group now subsumed into the Elders Council, applied social drama. Participants have used a repertoire of poems and sketches, put together in different ways, including developing a theatrical 'older people's kit' to demonstrate the difficulties older people face when they out and about in the city.

In terms of transforming Newcastle as a place, the Elders Council have worked with individual developers to include slip proof mats and improve health and safety. They have worked with city centre location to provide access to toilets, with Tesco's to develop a 'pensioner-friendly supermarket', including a scoping trip to Berlin to observe comparable German initiatives. In their studies on parks, the Council have called for more signage, benches and exercise equipment aimed at older people. The Council has started conversations on cohousing, holding meetings and producing briefings, alongside Newcastle City Council's initiatives Wellbeing for Life Strategy. The aim is to make Newcastle an age friendly city and 'to make Newcastle a great place to grow old'.

Northern Youth is a support organisation providing advice, information and development support services to clubs and groups. It is a partnership of over 100 youth clubs and projects – reaching over 8,000 young people a year. The organisation's main objectives are to promote the development of young people aged 8 to 25 in achieving their full physical, intellectual, social and spiritual potential by the establishment and development of clubs and groups throughout a county in the north of England. The organisation carries out the objects of a 'County Association' as defined in the byelaws of and in conformity with the principles and aims of Ambition and UK Youth.

Northern Youth have been closely involved in developing a scheme for discounted bus travel for under-19s in the county, and have raised funds and maintained momentum for the rebuilding of a local skate park, after the previous one was lost to floods. There have also been ongoing initiatives to connect policymakers and funders to hear young people's views and priorities. In 2013, the organisation co-ordinated a 'Young Persons Question time', while in 2012, Northern Youth was part of a pan-EU project called Community X-change. Participants travelled to Brussels to join 80 residents from rural areas in Hungary, Slovakia, Belgium and Ireland as part of the European Citizens' Panel, culminating presenting a set of recommendations for action to improve life in rural areas. Creative practice is an ongoing part of Northern Youth's work, and in 2012, when participants felt that they were not being listened to, they expressed their frustration by taking to the streets of Brussels, holding giant speech bubbles over their mouths, which said in both English and French: 'I travelled 874kms not to be heard'.

The third participant group is located in Stokes Croft, an area in Bristol, situated at the bottom of the A38, an arterial road which for many years was the main entrance to the city from the North, until the arrival of the M32 in the 1960s. Stokes Croft has a history as a site of nonconformity, lying beyond the original city boundaries with an active scene of theatres, dissenting schools and almshouses, particularly in the eighteenth century. The area has also been well studied by academics – for its grassroots arts cultural activism (Buser et al., 2013), as a site of performance (Karantonis, 2008), and as a test-bed for public archaeology with homeless people (Kiddey and Schofield, 2011). The 'riot' where a Tesco's store was damaged and looted has also been academically documented (Clement, 2012). Also central to the transformation of Stokes Croft are the Coexist CIC (who have an ongoing lease of the very large Hamilton House, from socially-minded property developers Connolly and Callaghan) as well as a wide range of independent traders, retailers and restaurateurs along this part of the A38 and in nearby Montpelier and St Paul's.

The research participants in Stokes Croft were representatives from the People's Republic of Stokes Croft. (PRSC). PRSC is a network of artists, performers and community engagers, incorporated as a Community Interest Company. It is facilitated by PRSC's Chairman Chris Chalkey, a social entrepreneur with ownership interests in key buildings here. Central to PRSC's mission is to seek to work with the skills of the whole Community including to 'promote and bring to fruition the notion of Stokes Croft as a Cultural Quarter, as a destination' (PRSC website, 2015). Closely involved with the development of the Stokes Croft Museum (now closed), PRSC have long promoted creativity in their work, reimagining and revisioning a urban landscape long characterised by many as derelict, with fluid, often-changing 'whimsy and artistry' (Buser et al. 2013: 608).

The research project findings included three headline messages from community participants: (1) the great appetite for positive action, working proactively to improve street furniture, buses, lavatory access and the aesthetic landscape; (2) the need to include many groups in society: the young, the older, immigrant groups and the homeless; and (3) that community participants need support from council officials to facilitate improvement and change and financial support to compare best practice and innovative designs.

Legal Consciousness in Creative Place-making

Legal consciousness research is often premised on these four schemas of 'before', 'with', 'against' and 'under the law' (Halliday, and Morgan, 2013). These heuristics, and there could be more, particularly in different cultures and in different times, are used to analyse how people engage with legal practices. The frameworks explore how 'dominant groups exercise power over subordinate groups' and explore how groups 'these accept their subordination through the influence about ideological ideas about their place in society', which Travers (2009, 75) has identified as the central theme in critical theory. In understanding hegemony, legal consciousness scholars may draw on Marxist thinking that social and/or economic class is the line of division, or they may premise race, sexuality or appearance (Abrego, 2011; Harding, 2006). Hegemony can come in many forms, and be intersectional (Harding, 2006), building on existing social hierarchies (Nielsen, 2000). For the purposes of this study then, law and legality are integral to understanding how spaces are created and recreated through social action.

Certainly, lawyers are fond of asserting that 'the law is all over' (Sarat, 1990), meaning that legal practice and provisions saturate every social experience. Yet as Levine and Mellema (2001: 205) remind us, 'always there' does not equal 'always important' to all participants. Legal provisions may determine who owns the land on which public spaces exist – and what the owner can do with that land – in that sense it is 'all over' as it is in the highways rules, the insurance provisions, the criminal law, the advertising restrictions. Still, legal practices and perceptions provide only a partial explanation of why places are the way they are – money, personality, status, age, experience, resources of all kinds – are all crucially significant as well.

In place-making, multiple factors, often quite mundane, can explain why particularly innovative approaches are used in one place but not another. In this project, for example, one of the most striking findings was the very short time people are considered to be young (broadly aged 16–18) in the work of Northern Youth, compared with how long we are old (aged from 50 onwards in the case of the Elders Council of Newcastle). Place-making is often a long-term project, taking years rather than months and draws on life experiences, which can make it more difficult for young people to engage in. Different levels in engagement in place-making are clearly not only about legal consciousness. Despite these caveats, however, there is much that can be understood about place-making action (and non-action) by understanding legal consciousness research.

In particular, what is hegemony in place-making? And how is it legally facilitated? One suggestion is that hegemony in place-making is the ability of a person or group in a position of power to determine the look, feel and use of a place. Often that will be the landowner, many are very tightly controlled by property owners (for example, 'retail quarters' or malls, Layard, 2010). Or it may be the local authority ('cleaning' off graffiti, for example); or police authorities using spatially divergent policing practice; or even hegemony by criminal gangs in some locations. The extent to which these are hegemonic practice will depend to some extent on how far others feel able to be involved and engaged.

Conventionally in English place-making, as in many other jurisdictions, involvement and participation is achieved through regulatory processes, particularly through planning at local authority (and now neighbourhood) scale. Neighbourhood planning is likely to be undertaken by groups that are willing to engage "with" the law. That have that specific legal consciousness. Strikingly, however, there are no regulatory procedures that enable the public to engage with property owners. Unless a landowner himself, or themselves if it is a corporate private or public owner, decides to engage with others on how a piece of land should look, feel, sound or smell, it is up to the landowner to decide (subject to planning, environmental and other regulatory constraints). This means that (even inadvertent) hegemony can occur quite easily unless there is some express facilitation for public involvement.

This is where this project differs slightly from other investigations into legal consciousness. It is concerned not with court cases or even 'problems' (as in Ewick and Silbey 1998) but rather a desire to get involved with place-making: to use the non-contentious regulatory processes of planning and place-making (assuming that they don't go to court) to make the city 'age friendly' or beautiful. All three of our groups started by trying to engage with these processes, assuming that these (legal) systems would give them a way in. It was only when they felt that they were not being listened to that they decided to diversify their strategies, pursuing other avenues.

Because these are generally not litigious processes, when Newcastle City Elders engage with place-making they engage with established local authority systems and practices. The aim is to gather evidence, which can be submitted into decision-making processes in order to 'make Newcastle a great city in which to grow old'. As one participant described it:

> We start by circulating a questionnaire to our [2000+] members and saying if we look at this what is important to you when you are going into the city centre – [doing a] similar exercise with parks and recreation areas and supermarkets. Build that evidence into a checklist for things to look for and then do a systematic assessment of those facilities to see how far they meet the expectations of older people or fail to meet them. That's the method that we use.

While in contentious (litigious) processes, contestation has been understood as playing a 'game', where one side would win, contesting in non-contentious procedures is slightly different. Regularly theorized and framed by planners, this form of participation almost always requires paperwork submitted through prescribed consultation processes. It was the sheer frustration with this process that led Bristol activists to operate differently ('under the law', see below). As one participant in Bristol put it:

> The way that the City Council want you to engage through talking shops – engaging with the processes – the only person sat round the table who's not being paid – if you could hook up all the hot air to a turbine you'd be able to solve the energy crisis and nothing happens. That is why I'm an advocate of direct action.

In Newcastle, however, the Elders Council were more prepared to engage. Beginning with a desire to demonstrate that the group were interested in the development of Newcastle, they became involved in the local strategic partnership. They now perceive themselves to be 'locked into a system' because the Elders Council is involved in the Quality of Life Partnership, which involves the City Council, the Elders Council and Age Concern Newcastle (now Age UK). Consequently, 'the main thing' this group have been working on is in the planning documents 'trying to make sure that we do influence the vision of the city, including a city for older people'.

This, then, is a clear example of a well-resourced (in multiple senses) group who have the time and the desire to engage with these consultative processes. Participants often received a pension, and so could afford to give up their time for free (though not all are volunteers). This was a very different situation than for the activists in Bristol, or the young people involved with Northern Youth. These everyday realities illustrate how legal consciousness intersects with social hierarchies. There are multiple reasons why people might be more likely to 'contest' than 'conform'. Where legal consciousness is useful, though, is to distinguish how different individuals and groups engage with (il)legality, to consider how this contributes to hegemony in place-making as well as challenges to hegemony (or neglect).

It also enables us to assess changes in consciousness over time. For what became evident through the project was that even the most positive, proactive participants can become disillusioned. As the Newcastle Elders explained: 'That is one route to get to the planners, but we need to step outside that system because your voice can get a bit muted if it's always going through another source. Have to make sure we do get our points across'. The participants in Newcastle described an involvement with a developer, noting how they felt that they had been used to 'soften up' the Council, so that when the application for planning permission went forward he could say that he had consulted with the Elders Council'.

Similarly, participants involved with Northern Youth were disappointed in the way that the 'Youth Parliament' was constituted. This was created to discuss the issues that young people across the county feel most strongly about: transport, employment, and leisure – and will be working on action plans over next few months together with Elected Members from each Borough as well as the County Council. Yet our participants felt that the way participants were selected, particularly from the public and grammar schools, rather than working with youth centres privileged some young voices over others.

The question for our participants then is. What to do when that disillusionment sets in, if the formal processes do not appear to hear your voices, if 'contesting' does not deliver? Here all three of our pioneer communities turned to creativity: creating a positive alternative. As one participant in Bristol put it: 'If you just rage about the council ignore you. If you create an aesthetic environment which attracts people because it makes them laugh, because it's beautiful, because it's bright and beautiful then people become engaged'. Sometimes, you are just making the point that you cannot participate in the process. This is what the young people from Northern Youth did. Having travelled to Brussels to take part in a consultative process, they portrayed their frustrations visibly, on street corners, creating an interactive art installation.

In the case of our participants in Newcastle in particular, then, creativity complemented engagement with established procedures where our participants used creative means to get around the conventional, legal, participatory processes. In Newcastle, for example, one participant explained how 'aesthetics matter: you can write a 50 page document but this won't have the same effect as a cartoon or the performance in front of the council. There is a visual impact with lots of pictures, creating colourful reports'. Certainly these creative interventions into the often rather technocratic and procedural planning processes are no panacea, as our Newcastle participants confirmed:

> Depending on which medium you're choosing, supposing you put it into a DVD the impact comes from actually showing it to groups and the impact will fade after a time. But if you have a report which you can go on referring to, you hope

> that will be longer lasting. Try to do both to make the main impact but also to have something available that can sustain it.

Nevertheless, as our participants confirmed, creativity then can be part of contestation, engaging 'with the law' by supplementing conventional documents with cartoons, films or theatre. In Newcastle, meanwhile, with the aging of members of the Old Spice drama group, the group is increasingly engaged in more conventional ways, 'contesting' by engaging with strategic decision-makers in the City Council as well as preparing their own research and lobbying documents. The type of creativity changes.

Such a supplemental, creative approach can also engage people not engaged in formal ways. As a participant from Northern Youth confirmed:

> Yes, video may not be long lasting. We are also doing graffiti art which will hopefully be photographed and taken wherever we go. [Once completed] … that message can be taken on so that when we're losing parts of money, with different perceptions of the Youth Services using a visual focus and drawing on projects young people are passionate about, including drama, music, photography and film can be useful.

Using these arts methods can be more effective than writing reports and interviewing, it 'keeps them on their feet' and particularly when young people are shy, they can feel more comfortable with an activity or acting in a film in a safe space but would not stand in front of councillors because it is intimidating. The participants from the Elders Council and Northern Youth always kept their creative practices within the law. They presented their interventions with consent, or on the pavement, or the highway. They did not break the law. Participants involved with both groups used creativity to supplement their engagement 'with' the law and established consultation processes.

A very different pattern of behaviour has emerged in Bristol, where participants' frustration with engaging in established legal processes to place-making was justified by engaging with what they saw positive, collective acts of dissent. If they could not engage with established processes, because planning hearings or property owners 'would not listen', then our participants decided to take matters into their own hands. Their approach is summed up by one piece of graffiti in Stokes Croft, 'When injustice becomes law, resistance becomes duty'. As the Chair of PRSC explained in Bristol:

> Law is as best a guideline … with regard to the law, you basically boil down to that philosophical question, is it right to obey an unjust law? Because the laws are generally made by those in power and they're generally made to suit them … the laws are made by the people who benefit, so property laws are made to benefit those who have property … everybody has to recognize the law – you wouldn't want to live without laws but there is a balance between obeying the

> letter of the law and taking a bit of a more cavalier attitude particularly when it's clear that [more] needs to be done.

This is a striking instance of a legal consciousness that can be characterised as 'under the law'.

Legal Geography in Creative Participation

If actions across all three case studies can, then, be variously framed as being 'before', 'with', 'against' and 'under' the law, how do these different characterisations relate to their spatial location? Just as legal consciousness demonstrates the reflexivity between the legal and the social, that cultural schemas frame understandings of, and approaches to, legality, so there is reflexivity between the spatial, the social and the legal. This is the central premise of legal geography. This matters because while participants 'on the ground' understand the systems, rules and practices that affect place-making, law *and legality* matter, even if they are not determinative.

Clearly, our participants from the three groups are working in very different social and spatial contexts and as legal geography scholars have often sought to remind us: place matters. As Pue (1990: 578), in an early clarion call for legal geography, proclaims: 'citizens, localities, and 'place' win when specificity is victorious'. Yet the legal rules, on property, planning and highways, for instance, are the same. They are general (within England), rather than locally specific. Certainly in each location legal practices differ: for the county where Northern Youth operates a key aim of local policymakers is to promote and facilitate tourism; in Newcastle, the legacy of T. Dan Smith still influences the design and use of the city centre today (as our participants frequently reminded us). In Bristol, meanwhile, attitudes to graffiti by both property owners and the police have evolved distinctively, with a more permissive approach being taken in Stokes Croft than in other locations.

So, as Pue (1990) suggests, specificity clearly matters. The application of legal rules in different sites will reflect the social and spatial contexts in which they are applied, even if they are nationally prescribed. Yet this is not as straightforward as it seems. Not all residents near Stokes Croft, for example, are happy with extent of the graffiti, and how it has spread into nearby neighbourhoods. This reflects Holder and Harrison's call for law to make room for local conditions and experience, and recognize the changing of laws to work in local contexts. There ask us to identify 'local legal universes' or 'legal localizations' – forms of regulation rooted in local conditions of existence' (Holder and Harrison 2003: 4). 'In turn', they continued, ' 'doing law' in geography … helps our understanding of how law shapes physical conditions and legitimates spatiality, and makes clear that law has a physical presence, or even many presences. This has the capacity to release law from its (imposed and self-imposed) confinement as 'word' (interpretation, meaning, discourse)' (Holder and Harrison, 2003, 5). We cannot explore place-making, without considering place, but this 'making' is shot through with law and legality.

Proceeding from legal geography enables us to understand that there is reflexivity here, both between social and legal place-making and between people and place. In legal consciousness research, these connections and understandings of 'second-order' layers of and 'people's perceptions about how others understand the law', matter. This is true both for local authority place-makers (running planning or highways consultation processes or liaising with property owners) and individuals who might dislike graffiti in 'their' neighbourhood. Working within this context requires an understanding of the spatial and social context within which:

> Legal knowledge prefigures or shapes subjectivity in that it provides a preexisting lens for experiencing meaning, for 'seeing' and thinking, for constructing and imagining in culturally 'sensible' ways. But subjects also actively sue legal knowledge instrumentally as a 'toolkit' of cultural resources, sorting and reconstructive from among variously acquired legal tools to understand social relations, to frame options, to assess relational or material consequences of various options, to formulate what is right and wrong etc. Legal consciousness is both thinking and doing. (McCann, 2006: xiv)

Unlike some of the other situations where legal consciousness 'raising' research has been done (Scheingold, 1974; McCann, 1994), there are no 'rights' to place-make (other than to engage in planning processes that our participants found so limited). By working 'under the law', prizing creativity, and beauty (however defined, and by whom) above restrictive legal rules can raise consciousness about capacity. Others can see the benefits (as they see them) to be gained from not asking first (for urban gardening, for example, or painting street furniture).

Nevertheless, place-making 'under the law', so to speak, can set a trend, as it has in Bristol. In 2010, Chris Chalkey, Director of the People's Republic of Stokes Croft was prosecuted and found guilty of causing criminal damage after painting a small sign that said 'Welcome to Stokes Croft, Cultural Quarter, Conservation Area, Outdoor Gallery' on the front of a gated community, 5102, at the bottom of Stokes Croft. He was fined £750; the cost of repainting the wall. This is a city where Banksy murals are regularly preserved, and the Council has considered holding public polls on whether individual street art paintings should stay or be painted over. Times have changed, however. One key witness for Chalkey at his trial, ex-RIBA President, George Ferguson, is currently Bristol's mayor, while Stokes Croft itself hosts competition for new pieces of street art and 'graffiti' tours are integral to the city's tourism strategy.

If we would like to see more creative place-making, then, there may then be something to be said for creative participants to be 'under the law'. This would not require advocating infringing property rights (clearly) but also not to wait for permission to intervene in a system that does not encourage creative participation.

Of course, artistic tastes can also be hegemonic, with some individuals gatekeepers of what is 'aesthetic' or 'beautiful' in individual neighbourhoods. As one commentator noted following Chalkey's prosecution (defending this particular notice and criticizing the prosecution) but saying:

> What I profoundly dislike is the idea that this is done for my benefit, mostly it is poor and ugly art and I don't see why I should have to look at it every day and feel grateful for a cool hip thing happening. Ugly is ugly after all, and there is plenty of that on offer.

Conclusion

This research then has asked why some participants engage in graffiti or beautification projects without explicit permission, while others feel unable to act. It has explored why some groups engage with developers through formal planning processes, while others give up. The answer suggested here is that in addition to different sets of resources, there are differences in legal consciousness. This is why the Newcastle Elders have continued, despite some frustration and no small amount of creative side-stepping, to engage with the non-contentious, consultative processes. While planning and place-making law is primarily non-adversarial, they have engaged in the process, 'the game', 'contesting' and working 'with the law'.

Conversely, as this chapter has explored, the young people involved with Northern Youth have been much less able to contest. Instead they have 'conformed', staying 'before the law'. It has also described how in Stokes Croft, in Bristol, there are clear instances of participants being 'under the law', of a collective belief in higher norms – of aesthetics, beautification and challenge, even if these overrode conventional property rights or planning processes. These are not singular characterisations but broadly capture the different approaches used by each set of 'creative pioneers' at the time of the study.

This leads us to ask whether we can use legal consciousness research to develop a theory of change. It seems that, while property, planning and highways rules are the same throughout the country, some groups and participants *do* feel able to 'place-make' more actively, while others feel unable to make change or participate in place-making. If this is true, and if we assume creative place-making to be desirable, how can we ensure that people become more involved in place-making and that creativity can flourish? In legal consciousness scholarship this question has conventionally been understood as raising 'rights' consciousness, particularly amongst low-income people, individuals from a racial minority (Albrego, 2011) or people with disabilities (Vanhala and Kelemen 2010). The suggestion then would be to make residents more aware of place-making processes (encouraging them to engage 'with the law'), while also explaining to decision-makers why mere opportunities to be consulted are insufficient if participants are 'before', 'against' or 'under' the law.

For ultimately, we cannot really expect many places to be like Stokes Croft, particularly in respect of the property ownership by key individuals in the locality. But perhaps we might think how to foster an 'under the law' legal consciousness and a 'can-do', DIY approach. We know that place-making is an 'everyday' process, happening over and over again, not limited to iconic projects. We need to understand the multiplicity, both between different people and between the interaction of people, place and law/legality.

This promotion of participants 'under the law', is of course profoundly different from the suggestion that 'communities need to place-make' for themselves. This is itself partially a hangover from the Big Society rhetoric, and partially the implication of initiatives in neighbourhood planning including the right to bid and the protection of community assets, certainly in more financially secure neighbourhoods under the Localism Act 2011. Arguably, under the law DIY creativity is less desirable than these more formal mechanisms, since it reprises the concern about who is 'the public', which is so often the case with DIY creativity and volunteers. For while some, particularly the creators of graffiti and (illegal) beautification projects may love their work, others may not, yet when participants operate 'under the law', there are clearly more limited channels through which broader input and discussion can go ahead. One person's graffiti can feel hegemonic to another. This is of course an inevitable consequence of allowing peoples 'passion for place' to be implemented without wider discussion. Perhaps it is simply the downside of resisting 'creativity by committee'.

Clearly then, when we think about place-making, legal provisions and legality matter enormously. In particular, understanding the legal consciousness of different groups can help us explain why some groups operate 'legitimately', as part of consultative processes or talking with property owners, while others operate 'illegitimately', wanting to turn the tide on what is considered creatively acceptable. There is no doubt that this has happened in Stokes Croft in Bristol, where the street art and beautification has become a key strategic aspect of Bristol's tourism and regeneration strategy. Trying to understand creativity or place-making without understanding the legal provisions that are 'all over' these activities and how practices and events are influenced by legal consciousness, will produce only a partial analysis. Law and legality matter. And when people meet place, they bring their legal consciousnesses, as with all their other social and cultural framings and resources, with them.

Annex: Policy Proposals from the Creative Participation Project

1. Communities should be encouraged and supported to engage both in conventional consultation processes and through more creative mechanisms, including theatre, film, cartoons, music and art;

2. Both elected representatives and employees in local government should work with communities proactively. Officers should not become a hidden hurdle to reform;
3. Planning use classes should differentiate between socially and culturally beneficial uses and those that pursue solely economic aims. Local, place-based concerns should be capable of being 'material considerations' under planning law;
4. Outline planning permission should incorporate some commitments to principles of internal design rather than focusing solely on the size and location of the development;
5. Criminal laws and planning rules should not be used to stop bottom-up improvement of areas. Local communities should be able to adorn their environments – through planting, art and signage – without the consent of the local authority;
6. Support, both financial and logistical, should be made available to enable communities to see examples of good practice elsewhere;
7. The new neighbourhood planning proposals need to engage with the multiplicity of communities. There are many publics and communities that all need to be engaged, including more marginalised groups. This will require time and resources. Bottom-up initiatives need support if they are not to become simply a base for the most vocal;
8. Conventionally unrepresented groups, particularly socially and economically disadvantaged young people, can be engaged through more creative mechanisms such as music and art. This builds individual confidence and strengthens links with the local community and other related groups;
9. Communities often become most engaged in positive projects rather than critiques of existing policies. This requires some funding as well as access to expertise. It is more productive than a negative critique of planning applications and the scrutiny of planning documents.

Acknowledgements

This research was funded by the AHRC (AH/J501553/1) and undertaken while Antonia Layard was working on a Mid-Career Fellowship on *Law, Localism and Governance*, funded by the ESRC (ES/J004642/1). We are very grateful for both Councils' for their support throughout the project. We would like also like to thank the editors and contributors to this volume for their thoughtful comments and discussion, which have undoubtedly strengthened the chapter. More information about the project can be found at creativeparticipation.com.

References

Abrego, L. (2011) Legal consciousness of undocumented Latinos: fear and stigma as barriers to claims-making for first- and 1.5-generation immigrants. *Law and Society Review* 45(2), 337–370.

Aitken, D. (2012) Trust and participation in urban regeneration. *People, Place and Policy Online* 6(3), 133–147.

Bartel, R., Graham, N., Jackson, S., Prior, J., Robinson, D., Sherval, M. and Williams, S. (2013) Legal geography: An Australian perspective. *Geographical Research* 51(4), 339–353.

Bennett, L. and Layard, A. (forthcoming) 'Legal Geography: Becoming Spatial Detectives'.

Blomley, N. (1994) *Law, Space, and the Geographies of Power*. New York: Guilford Press.

Blomley, N. (2003) 'From 'What?' to 'So what?': Legal Geography in Retrospect', in Holder, J. and Harrison, C. E. (eds) *Law and Geography*. Oxford: OUP.

Blomley, N. (2004) *Unsettling the City: Urban Land and the Politics of Property*. London: Routledge.

Blomley, N. (2007) Making private property: enclosure, common right and the work of hedges. *Rural History* 18(1), 1–21.

Braverman, I., Blomley, N., Delaney, D. and Kedar, A. (eds.) (2014) *The Expanding Spaces of Law: A Timely Legal Geography* Redwood City, CA: Stanford UP.

Buser, M., Bonura, C., Fannin, M. and Boyer, K. (2013) Cultural activism and the politics of place-making. *City* 17(5), 606–627.

Clement, M. (2012) Rage against the market: Bristol's Tesco riot. *Race & Class* 53(3), 81–90.

Cochrane, A. (1986) Community politics and democracy, in Held, D. and Pollitt, C. (eds) *New Forms of Democracy*. London: Sage, 51–72.

Conley, J. and O'Barr, W. (1990) *Rules versus Relationships: The Ethnography of Legal Discourse*. Chicago: Chicago University Press.

Cotterrell, R. (1997) *Law's Community: Legal Theory in Sociological Perspective*. Oxford: OUP.

Cresswell, T. (1996) *In Place-out of Place: Geography, Ideology, and Transgression*. Minneapolis: University of Minnesota Press.

Delaney, D. (2010) *The Spatial, the Legal and the Pragmatics of World-making: Nomospheric Investigations*. London: Routledge.

Ewick, P. and Silbey, S. (1998) *The Common Place of Law: Stories from Everyday Life*. Chicago: Chicago University Press.

Fritsvold, E. (2009) Under the law: legal consciousness and radical environmental activism. *Law & Social Inquiry* 34(4), 799–824.

Galanter, M. (1974) Why the' haves' come out ahead: Speculations on the limits of legal change. *Law and Society Review* 9(1), 95–160.

Graham, N. (2010) *Lawscape* London: Routledge.

Halliday, S. and Morgan, B. (2013) I Fought the law and the law won? legal consciousness and the critical imagination. *Current Legal Problems* 66(1), 1–32.

Harding, R. (2006) Dogs are 'registered', people shouldn't be: legal consciousness and lesbian and gay rights. *Social & Legal Studies* 15(4), 511–533.

Holder, J. and Harrison, C. (2003) *Law and Geography*. Oxford: OUP.

Karantonis, P. (2008) 'Leave no trace': the art of wasted space – the People's Republic of Stokes Croft. *Performance Paradigm* no. 4, no pagination.

Katz, L. (2008) Exclusion and exclusivity in property law. *University of Toronto Law Journal* 58(3), 275–315.

Kennedy, D. (1980) Toward an historical understanding of legal consciousness: the case of classical legal thought in America, 1850–1940. *Research in Law and Sociology* 3(1), 3–24.

Kiddey, R. and Schofield, J. (2011) Embrace the margins: adventures in archaeology and homelessness. *Public Archaeology* 10(1), 4–22.

Klare, K. (1978) Judicial deradicalization of the Wagner Act and the origins of modern legal consciousness, 1937–1941. *Minnesota Law Review* 62(3), 265–340.

Kress, K. (1989) Legal indeterminacy. *California Law Review* 77(2), 283–337.

Kritzer, H. and Silbey, S. (eds) (2003) *In Litigation: Do the Haves Still Come Out Ahead?* Redwood City, CA: Stanford University Press.

Layard, A. (2010) Shopping in the public realm: a law of place. *Journal of Law and Society* 37(3), 412–441.

Layard, A. (2014) Freedom of expression and spatial (imaginations of) justice, in D. Kochenov (ed.) *Europe's Justice Deficit?* Oxford: Hart Publishing.

Levine, K. and Mellema, V. (2001) Strategizing the street: how law matters in the lives of women in the street-level drug economy. *Law & Social Inquiry* 26(1), 169–207.

Litowitz, D. (2000) Gramsci, hegemony, and the law. *BYU Law Review* 2(1), 515–551.

Matthey, L., Felli, R. and Mager, C. (2013) 'We do have space in Lausanne. We have a large cemetery': the non-controversy of a non-existent Muslim burial ground. *Social & Cultural Geography* 14(4), 428–445.

McCann, M. (1994) *Rights at Work: Pay Equity Reform and the Politics of Legal Mobilization*. Chicago: University of Chicago Press.

McCann, M. (2006) On legal rights consciousness: a challenging analytical tradition, in Fleury-Steiner, B., and Nielsen, L. B. (eds) *The New Civil Rights Research: A Constitutive Approach*. Farnham: Ashgate.

Merry, S. (1990) *Getting Justice and Getting Even: Legal Consciousness among Working-Class Americans*. Chicago: University of Chicago Press.

Mezey, N. (2001) Out of the ordinary: law, power, culture and the commonplace. *Law & Social Inquiry* 26(1), 145–167.

Mitchell, D. (2003) *The Right to the City: Social Justice and the Fight for Public Space*. New York: Guilford Press.

Nielsen, L. (2000) Situating legal consciousness: experiences and attitudes of ordinary citizens about law and street harassment. *Law & Society Review* 34(4), 1055–1090.

People's Republic of Stokes Croft. (2015) http://www.prsc.org.uk accessed 15 February 2015.

Philippopoulos-Mihalopoulos, A. (2007) In the lawscape, in Philippopoulos-Mihalopoulos, A (ed.) *Law and the City*. Abingdon: Routledge-Cavendish.

Philippopoulos-Mihalopoulos, A. (2011) Law's spatial turn: geography, justice and a certain fear of space. *Law, Culture and the Humanities* 7(2), 187–202.

Pue, W. (1990) Wrestling with law: (Geographical) specificity vs. (legal) abstraction. *Urban Geography* 11(6), 566–585.

Pollock, V. and Sharp, J. (2012) Real participation or the tyranny of participatory practice? Public art and community involvement in the regeneration of the Raploch, Scotland. *Urban Studies* 49(14), 3063–3079.

Sarat, A. and Felstiner, W. (1986) Law and strategy in the divorce lawyer's office. *Law and Society Review* 20(1), 93–134.

Sarat, A. (1990) … The Law is all over: power, resistance and the legal consciousness of the welfare poor. *Yale Journal of Law & the Humanities* 2(2), 343–379.

Scheingold, S. (1974) *The Politics of Rights: Lawyers, Public Policy, and Political Change*. New Haven: Yale UP.

Silbey, S. (2005) After legal consciousness. *Annual Review of Law and Social Science* 1, 323–368.

Silbey, S. (2010) Legal culture and cultures of legality, in Hall, J., Grindstaff, L. and Lo, M. (eds) *Handbook of cultural sociology*. Oxford: Routledge, 470–479.

Travers, M. (2009) *Understanding Law and Society*. Oxford: Routledge.

Vanhala, L. and Kelemen, R. (2010) The shift to the rights model of disability in the EU and Canada. *Regional and Federal Studies* 20(1), 1–18.

Warren, S. (2014) 'I want this place to thrive': volunteering, co-production and creative labour. *Area* 46(3), 278–284.

PART II
Policy Connections, Creative Practice

Chapter 6

Bridging Gaps and Localising Neighbourhood Provision: Reflections on Cultural Co-design and Co-production

Ginnie Wollaston and Roxanna Collins

Introduction

Birmingham City Council (BCC) administers a city of 1,085,400 residents (BCC, 2014a), divided into 10 districts of approximately 100,000 residents, each district containing four wards. Birmingham City Council Culture Commissioning Service (CCS) are part of the Economy Directorate and fall within the Culture & Visitor Economy Service. Birmingham's 2010–15 Cultural Strategy (Birmingham Cultural Partnership, 2010: 4) outlines the multi-faceted benefits of participation in arts and culture:

> Participation in culture is inherently a good thing – it challenges perceptions, prompts feelings of happiness, sadness, anger and excitement, creates moments of personal reflection and enables people to understand the world they live in, its possibilities and the cultures of others more profoundly. Cultural activities can also deliver a range of other outcomes including health and wellbeing, social and community cohesion, civic engagement, economic impact, development of transferable skills and improved environment.

However, two reports prior to this strategy (Ecotec, 2010; Vector Research, 2011) revealed that many areas in the outer districts of Birmingham had poor levels of participation in culture and neighbourhood activities. Five main barriers to participation were identified: high cost of participation; lack of awareness of amenities and attractions; perceived cultural elitism; public transport deficiencies and issues of distance (as many of Birmingham's cultural venues are based within the city centre); and community safety concerns – in some areas people do not feel safe in their neighbourhood and would not look to engage with local activities (Ecotec, 2010).

Follow-up research into cultural participation concluded that 'individual non-participants and the neighbourhoods in which they lived […] may not actually be culture-ready' as in some cases neighbourhoods were perceived as 'cultural deserts' (Vector Research, 2011: 7). It was recommended in the report that 'intervention should be initially at a very basic level and addressed towards the people themselves and their neighbourhoods' (ibid.: 7). The recognition that people and local neighbourhoods could be the solution became embedded in the new cultural policy.

Birmingham's Cultural Strategy (Birmingham Cultural Partnership, 2010) revised the Cultural policy around four key themes, in particular the 'Culture on Your Doorstep' strand taking cultural activity direct to local neighbourhoods. The strategy recognised low levels of participation and engagement, and in two action points recommended 'improving communication of full range of opportunities for cultural participation at neighbourhood level' and 'embedding culture in new models of place based budgeting, coproduction and shared services' (Birmingham Cultural Partnership, 2010: 10).

While Birmingham's Cultural Strategy highlights participation as the key activity required to create an intervention in neighbourhoods to overcome barriers – the methodology of cultural co-design created additional interventions to underpin and involve residents in co-production. A variety of projects proposed and delivered with resident groups involved residents in: learning fund-raising skills; writing briefs for artists; being part of steering groups that managed projects and budgets; and delivering arts projects in their neighbourhoods, mentored by local arts co-ordinators. In some areas these steering groups brought together new partnerships between education, health, local business and resident associations. Events run by these partnerships included arts activities within other community events. This required planning that addressed geographical and social isolation, so that selected locations were accessible by public transport or within walking distance for local residents. Given that most arts venues are city centre based this meant using non-traditional sites for activities and using innovative marketing and community engagement strategies to persuade audiences to attend new locations. The Cultural Pilot programme (2012–14) was overseen by the CCS in three targeted neighbourhoods (based on the community budget pilot areas) to: increase participation in local arts; strengthen local cultural infrastructure; and empower residents through co-designing and co-producing arts and cultural activities. An additional aim was to pilot new sustainable business models and foster creative community renewal.

These aims translated into four pilot outcomes: increased levels of local resident participation in arts and cultural activity; residents involved in co-design and co-production of local arts and cultural activity; local infrastructure for arts and culture strengthened; and local residents' increased capacity to engage in neighbourhood activities.

The Cultural Pilot approach built on existing structures established by the CCS, namely the Local Arts Fora and Arts Champions. The Arts Champion programme (set up in 2005) requires each of the 12 large-scale arts revenue organisations of the city to take a strategic role within one District of the city over a three year period. The principles of the scheme are to provide local activity for residents in the District and to link this activity to the particular cultural offer of that organisation through different annual projects. To amplify this focus on the local provision, the Arts Champions programme (BCC, 2013a) became more directed towards working with the Local Arts Fora. These organisations, mostly based in the city centre, include the Birmingham Repertory Theatre, Birmingham Royal Ballet, Birmingham Hippodrome, Birmingham Opera Company, City of Birmingham Symphony Orchestra, DanceXchange, Ex Cathedra, IKON Gallery, Midlands Arts Centre, Sampad, The Drum and Town Hall Symphony Hall. The programme varies from offering participatory opportunities led by professionals to producing performance or exhibition materials to present take part in local festivals and events.

The Arts Champions also work with the relevant Local Arts Forum to build a dynamic network to bring together artists, community groups, BCC district officers and the CCS. Each Local Arts Forum is led by a small arts organisation or individual and they are contracted to: develop district arts plans; hold local contacts databases; provide audits of arts activities in local venues; develop active web and social media profiles and deliver local events to increase participation. Over three years, they have developed diverse and different networks of artists, community and youth groups, venues and district personnel across the city.

However, the larger strategic role in developing arts and cultural activities that might engage with approximately 100,000 residents across four wards within each district is hampered by limited resources and time. Successful models of operation have been the result of collaboration and partnerships between artists, local venues and libraries, partner organisations such as resident associations, schools and colleges, businesses and shopping centres.

This chapter critically analyses the extent to which the three Cultural Pilots achieved their anticipated outcomes. In particular, the success of cultural co-design and co-production will be reviewed in light of the different approaches taken in each pilot area, and how they have built on existing infrastructure to bridge gaps and localise neighbourhood provision.

Cultural Pilots: Area Identification and Focus

The Localism Act (DCLG, 2011) was integrated into the BCC Leader's Policy Statement in June 2012 following the change of local leadership (BCC, 2012a: 14). In September 2012 Birmingham was selected as one of ten neighbourhood level Community Based Budget pilots co-ordinated by the Department for Communities and Local Government (DCLG, 2013; Local Government Association and DCLG, 2013). The Birmingham vision was to develop:

> A 'triple devolution' approach that includes the development of a strong city region to drive investment and economic growth, a new model of integrated public services at the city level and strong communities engaged with the public services through neighbourhood community budgeting. We are clear that the

> neighbourhood is a crucial focus for the redesign of local public services because it is at this level that services can come together to invest in prevention – whether it be working with troubled families, engaging the community in preventing environmental problems, undertaking minor housing repairs before they become big ones or helping people to live healthier lives. (BCC, 2012b)

The three *Community Based Budget* areas were Balsall Heath (part of Sparkbrook ward); Castle Vale (part of Tyburn ward) and Shard End (a ward in Hodge Hill District). Some of the Community Based Budget Pilot aims were carried forward into the Cultural Pilots. This included: resident-led change with residents playing a more distinctive role in service redesign; strengthening the partnership between residents and statutory bodies; enabling residents, families and communities to become more resilient and solve their own problems; sharing learning between the three Birmingham neighbourhoods and pooling budgets and resources where appropriate at a local level so that better outcomes can be achieved.

The three Pilot areas contrasted in geography and socio-economic levels, housing stock, ethnicity of population, levels of social capital, as well as community and arts infrastructure (venues and levels of activity). It was decided to run the Cultural Pilot in the same areas, modelled on the neighbourhood approach to Community Based Budgeting (Local Government Association and DCLG, 2013) but with some differences. The model using arts and culture needed to: be flexible in outcomes delivery; be managed centrally through the Culture Commissioning Service; embrace Local Arts Fora and Arts Champions programme; and test out new ways of engaging with residents through cultural co-design and co-production. Funding from Arts Council England (ACE) and BCC (combined £95,000) supported the first phase of the Cultural Pilots to test out the methodology and delivery from December 2012 to March 2014.

Balsall Heath is on the eastern side of the city in Hall Green district, bordering the famous 'Balti Triangle' of restaurants. The District Strategic Assessment shows the area has one of the most ethnically diverse and transient populations in Birmingham of 14,000 residents. Multiple deprivation has been a constant reality (BCC, 2014b), and levels of crime (in particular, prostitution) ran high some 30 years ago. Owing to community-led engagement and positive action led by the Balsall Heath Forum and St. Paul's Development Trust, this area now has one of the highest percentage of people feeling proud to live in their area (BCC, 2014c). The annual carnival has provided a focus of community engagement in cultural activity over 30 years but there is also a growing number of independent artists drawn to the Old Print Works (OPW), a hub for creative ventures, artists and makers, to develop a range of arts activities through workshops or performance. The theme for the Community Based Budget focused on the environmental agenda to clean and green the area, and further build on neighbourhood planning.

Castle Vale is an estate built on the old aerodrome where World War II Spitfires were tested. Following resident action, post-war high-rise were pulled down in the 1990s and 2000s and new housing developed. The regeneration of the area left a legacy of positive community engagement in housing stock, community services and health care. The District Strategic Assessment for Castle Vale forms part of Tyburn ward within Erdington district, with approximately 12,000 residents (BCC, 2014b). The Community Based Community Budget process focused on health and well-being while the 21st anniversary since the work began to regenerate the estate provided a focus for the Cultural Pilot.

Shard End is the largest area covered by a Cultural Pilot, comprising an entire ward of 26,000 residents within the Hodge Hill District. It is better known to local residents as a collection of five villages including Tile Cross and Lea Village. The area has poor local transport infrastructure, and District Strategic Assessments show deprivation statistics including long term unemployment have not changed over the past 10 years (BCC, 2014b). The Community Based Budget pilot focused on families with complex needs and worklessness. In terms of arts participation, Shard End has been referred to as a 'cultural desert' (Birmingham Cultural Partnership, 2011: 16). There are several parks, and the River Cole running through the central Kingfisher Park. Recent new buildings such as the Shard library and the Pump youth centre, (hosting the only arts organisation, Reel Access) have provided new opportunities for arts and community initiatives.

Theories Influencing Cultural Pilot Development

One of the most important elements of the Cultural Pilot approach was the shift from an arts participation-only model, to a model that supported residents and artists to co-design and co-produce cultural and arts activities themselves. This approach meant a change in the way that artists might work as collaborators with residents and communities. Gillespie and Hughes' (2011) *C2 Positively Local* engagement methodology for community change (originally used in social care practices) was adapted as an open and flexible action research framework to find key people, establish a shared vision to help steer the arts project from beginning to completion and to identify and implement change in an area. It provided a seven step method for finding the energy for change and transforming that into action informed by and with residents.

The research required investigation into existing levels of participation by residents in cultural provision. At the beginning of establishing the Local Arts Fora and Cultural Pilots, the issue of sustainability was raised and embedded into the research briefs. The Cultural Pilot was not envisaged as a project which would come and go, but as a way of trialling a different approach to co-creating cultural provision collaboratively between local residents and other area stakeholders. This would require new models of engaging with individuals and community groups, new partnerships and potentially developing new business models that could emerge and sustain these activities into the future.

The community arts ethos of equality, inclusivity of access and participation is a recognised successful model. The intrinsic value of participation in arts and culture has been widely acknowledged, in particular its ability to increase participants' skills and confidence, support them to find their voice and to be heard (Webster, 1995; Clements, 2004; Matarasso, 1997). The process and outcome of an arts activity may not only be inspirational and a positive creative experience in itself, but also is a valued tool for community development, sharing skills, bringing people together and allowing resident ownership of projects and places. Arts practitioners Felicity Allen and Malcolm Dickenson (1995: 18) state:

> Community art is the manifestation of an ideology. What makes it different from public art or art in the community, is its long-term cultural and political ambition […] revolving around the notion of empowerment through participation in the creative process.

The *C2* approach (Gillespie and Hughes, 2011) was introduced to the initial research period as a mechanism for empowerment and ownership of process, and remained a useful method adapted to arts development in line with the community arts ethos, particularly in areas with low engagement and limited resources. The 'Seven Step Model' identified a process of consultation aimed at effecting positive change in the community. When adapted to an arts development process the main steps were followed but alongside this, important indicators for arts development were identified. In the first step – 'locate energy for change' – appropriate individuals were found to join steering groups and lead projects. For step two – 'create positive vision' – a theme or artistic project was found that held local resonance around more than one individual's artistic goal. This required group agreement to the project, and agreement with the Local Arts Forum. Step three –'listen to the community' – meant going back to stakeholders within the community to agree on the artistic vision and, step four, was to bring key members to formalise the partnership and undertake the implementation of that vision. Step five, 'sustain momentum', relied on the role of the co-ordinator or project lead, to facilitate the group to 'take action' (step six). Finally the interesting and flexible step seven ('renew and adapt') allowed artistic realisation and learning to recognise when and how to adapt the project to the available resources and abilities of the group as well as more general practicalities.

If appropriately applied, the model is intended to create a shared and realistic vision out of initial research, consultation and artistic ideas. The learning came when this model was not followed or applied and an individual steered action without taking the collective views of residents with them. In the Shard End report, some resident feedback was that they felt 'over consulted' as the Community Based Budget pilot was happening at the same time. They felt that consultation with no visible action or outcome to show for this was frustrating. The *C2* methodology 'supports residents to reclaim their capacity to direct change for themselves' (Gillespie and Hughes, 2011: 43) and this was directed into co-design and co-creation of arts activities.

The Cultural Pilots had time-limited funding so it was important to determine phases for research, consultation and delivery. CCS coordinated the phased project plan and evaluation across the three areas.

In Phase One, the operational framework was agreed with funders (ACE and BCC) and the BCC Community Based Budget leads in the three areas. The aims, outcomes and operational plan were managed centrally through CCS. A brief went out to tender and three co-ordinators were appointed in December 2012 to undertake the initial *C2* research from January to March 2013.

Phase Two involved the coordinators identifying the priorities for the neighbourhood, meeting with key stakeholders to establish ideas and priorities for engagement, an audit of current cultural provision in venues including arts organisations and individuals and some assessment of the needs and the gaps in provision. Between the three co-ordinators they held face-to-face interviews with 72 individuals (including councillors, artists, community and business leaders) and consulted with 96 groups potentially interested in arts activities, or partnering with arts and cultural activities in the neighbourhood. Two out of three co-ordinators used the *C2* seven step process to identify new voices who might want to engage in cultural or arts activities and new partners who could help to deliver activities. One coordinator identified a particular proposal based on their research that was weighted towards one aspect of the neighbourhood and the business community. This resulted in action that took time and energy but did not result in a shared vision, proving difficult to realise steps four to seven of the methodology. This was a valuable lesson as to what worked well and what the challenges and barriers can be when engaging partners and individuals in a given community. The learning from this led to a slightly revised delivery methodology for this area.

A report from each co-ordinator was submitted at the end of Phase Two to an external panel who then decided how the proposals could be funded within the approximate allocation of £20,000 per neighbourhood. Phase Three involved the delivery of arts projects in two delivery plans from May to September 2013 and from October 2013 to February 2014. Contracts with briefs were awarded to individual Delivery agencies (arts or community organisations) against confirmed proposals, timelines and budgets. Each Cultural Pilot was required to undertake a presentation of a case study as part of a final Symposium 'Cultivating Culture', which was held in the Library of Birmingham in March 2014 (BCC, 2014c). Three Local Arts Co-ordinators oversaw the delivery of the work during Phase Three and reported to CCS managers of each project.

An external evaluation team (Merida Associates) were appointed halfway through the project. They devised an outcomes evaluation framework to evidence against indicators from feedback during the second delivery phase. A methodology of questionnaires, surveys, individual and group feedback provided evidence from which to evaluate the model and this approach.

The final evaluation report *Igniting a Spark* (Merida, 2014) identifies the learning from these Cultural Pilots, with strong recommendations to: continue this way of working with the *C2* model; appoint skilled arts-based coordinators;

develop evidence based commissioning with better data capture and monitoring tools; share the lessons from the pilots; and build sustainability through better city-wide policy alignment.

From Shared Visions to Shared Delivery

In Balsall Heath, an element of Pilot arts activity (£13,000) was undertaken by St. Paul's Community Development Trust (also coordinators of Hall Green Local Arts Forum), expanding on existing projects such as the Carnival or developing partnerships with new or active groups. Combining the 2013 Carnival with the Britain in Bloom competition brought record audiences of 5,000 people to see the themed butterflies, bees, bugs and bloom procession. Two new groups of young people were supported by the Pilot to engage with the Carnival for the first time; young people and families were invited from a local Play Centre in a neighbouring park and from a local Primary school bringing an additional 55 children and 11 adults into the procession. The Drum (Hall Green District Arts Champions) supported the music programme in the park at the end of the event bringing beneficial cultural and environmental agendas together. Additional activities facilitated by St. Paul's involved increased profile for the arts in the Balsall Heathan newspaper (free to all residents), facilitation of a new Carnival of Light and Haunted Circus event in St Paul's venue, and the creation of a Ladypool Road Heritage Trail with the Still Walking Festival and Balsall Heath Local History Society (CCS, 2014a).

The remaining £7,000 was shared between three independent artists and arts organisations based at OPW, appointed through an open tendering process. Some Cities Community Interest Company ran workshop sessions for young people developing skills in photographic portraiture, street photography and reportage (CCS, 2014b). Young people were encouraged to use poetry and prose to present their images in a final public sharing. The Ort Café (based in OPW) worked with a local partner from the Somalian community to produce and manage a Somalian music event hosted in the OPW venue (CCS, 2013a). Two artists developed a new partnership between the local Play Centre and OPW, co-creating the story and shadow puppet presentation of 'Wishella' with 32 young children and adults in the OPW (CCS, 2014c).

In Castle Vale the proposed funding was divided into two projects: £15,000 was allocated to develop a three day Castle Vale Festival (held in a circus top on the local playing field) in September 2013 to celebrate 21 years since regeneration transformed the estate. The festival was planned and delivered by a 20 strong steering group, facilitated by the Cultural Pilot co-ordinators Active Arts, a local arts organisation with a history of delivering creative participatory projects in the area since 2008. The festival included an opening ceremony involving over 90 residents (from 7 years to 86 years), a Family Fun day, local tours around the estate, two evenings of music ('Live and Loud', a 'Big Sing') and a contribution from Town Hall Symphony Hall (the district's Art Champions), and a volunteer programme to manage the event. A total audience and participation figure of 4,800 attended these events (Active Arts, 2013a).

A smaller scale project (£5,000) was set up to pilot a GP Social Prescribing Project to offer three types of artistic activity over an eight week course for those with long term medical conditions. Individuals were either referred by local GPs or self-referred onto the project, delivered in partnership with the Castle Vale Community Regeneration Services (CVCRS) in the local community health centre, The Sanctuary. Referrals were made to a range of sessions including singing, photography, storytelling and creative writing. The project involved 27 adults (20 female; 7 male) engaging 19 adults with long term conditions and eight with disabilities (Active Arts, 2014).

In Shard End, the delivery of the arts projects was shared between Reel Access (also the Hodge Hill Arts Forum coordinator and Shard End Cultural Pilot coordinating organisation); the Yorkswood Resident Association and DiVaS (a women's domestic abuse awareness and support group). The initial project managed and delivered by Reel Access (£13,000) included bringing an outdoor theatre production of 'Alice – an Extraordinary Adventure' produced by Heartbreak Productions into Kingfisher Park. This production involved 10 young children (aged 8 to 10 years), who were coached to participate in the song and dance routines by a local youth theatre leader (CCS, 2013b).

An accredited training programme (led by Alkhami youth and community development company) gave 15 unemployed adults the opportunity to event manage the production in the park and gain an accredited Security Industry Authority (SIA) security badge qualification to ensure future employment. A positive partnership with Heartbreak Productions, the local Parks Service and personnel from the Shard library allowed the professional and community cast to use the Pavilions changing rooms in the park in partnership with the Friends of Kingfisher Park and to run two preparatory workshops at the Shard with a local box office run by staff. This drew a local audience of over 200 into the playing fields, most of whom had never seen live outdoor theatre in this part of the park that previously had only been used for cricket (CCS, 2013b).

Reel Access further developed new partnerships with the local Family Centre to devise a creative programme (Handprints) using drama role play and visual arts to offer the opportunity for fathers to play with their sons and create positive relationships through engagement. They supported the local Yorkswood Resident Association to take on the contract (£4,000) to restore a stone Griffin on their estate and run a Griffins Day event to encourage resident participation and engagement. This has led to a Heritage Lottery bid to restore the remaining five Griffins sculptures in partnership with the local Primary School. Similarly Reel Access facilitated the partnership development with DiVas with professional arts organisation Women and Theatre, to undertake a drama programme (£3,000). These local organisations embraced under-represented groups, especially families with complex needs and those affected by worklessness (CCS, 2013b).

Increasing Participation in Arts and Cultural Activities

The Pilots all successfully achieved the first of the Pilot's outcomes: increased participation in local arts and cultural activities. Balsall Heath Carnival engaged an estimated 5,000 audiences and 200 participants (an increase of 50 participants from the previous year). A participant commented:

> It was the biggest, brightest and most beautiful Carnival ever. Everyone should be proud that we run such fantastic events in Balsall Heath, and fingers crossed for a Britain in Bloom 'Gold Award', because there can't be any community with so many hundreds of human helpers bringing forth blooms and benign bugs. (CCS, 2013a)

The first Castle Vale Festival built on previous experiences of running events such as the Community games event in 2012. An increase in audiences and participants rose from 2,700 to 4,800. The Festival event over three days was also managed by 38 volunteers and provided sustained activity in partnership involving sports and cultural activities (Active Arts, 2013a). In Shard End, the approach to develop one big event in partnership with Heartbreak Productions and smaller projects with local organisations targeting particular communities engaged residents for the first time as participants, audiences and co-creators. The total participation and audience engagement was small in comparison to the other Cultural Pilots but as a starting point of engagement the results were very positive.

One of the key differences the Pilot made to these events was to engage residents in arts for the first time (67 per cent in Castle Vale, 90 per cent at Balsall Heath's Lantern Festival, and 43 per cent at Shard End's Griffins Day; in Merida, 2014: 16). The increase in the depth of engagement, allowing residents to plan and steer an event, volunteer as event stewards or event organisers or to lead on programming cultural events was a huge step in taking responsibility for managing the event. Those engaged in this way fully recognised the intrinsic and instrumental value of arts and cultural activity and overcome the barrier of 'perceived cultural elitism' (Ecotec, 2010: 35). As one resident involved in Castle Vale's 'Big Sing' event commented:

> Maybe before I joined the choir I might have had a different view and thought it's not really for me, but since I've joined the choir I've gone out and done things in public I would have never dreamed of doing before. (Merida, 2014: 23)

Across the three Pilots there were a total of 202 volunteers involved in a range of roles from steering groups, event organisation, catering, programming, to community marketing and audience development. Across the three Pilots there were a total of 10,119 audience members attending events, performances and festivals and 1,330 participants involved with projects in workshops, arts and cultural activities or performances (Merida, 2014:16).

Across all the Cultural Pilots under-represented groups were involved for the first time (DiVas and Yorkswood Resident Association in Shard End), individuals with long-term medical conditions in Castle Vale and culturally diverse groups in Balsall Heath. Through the decision to put out an independent tender brief to small arts organisations in Balsall Heath an additional 144 participants; 148 audiences; 4,000 Twitter followers and 2,750 Facebook contacts were created. The diversity of population included individuals from BME communities who identified themselves as Black British, African (Somalian or named African countries), Asian (Indian, Sikh and Pakistani) or British Asian, European, Chinese, and Chilean. The three organisations have a workshop space in OPW, which has created a positive hub of artistic activity to engage with local residents in this culturally diverse neighbourhood (CCS, 2013a).

Resident Co-design and Co-production

The second of the Pilot's outcomes was for residents to participate as co-designers and co-producers of local arts and cultural activity. A shared understanding of the meaning of 'participation' is central to genuine co-production and co-design engagement, as guided by Arnstein's (1969) Ladder of Participation and in the *C2* methodology (Gillespie and Hughes, 2011). The City Council's *Creative Future II: A Strategy for Children, Young People and the Arts* (BCC, 2009) distinguishes the roles of audience members, participants, creators and leaders, and these definitions have to some extent been applied to *Culture on Your Doorstep* policy (Birmingham Cultural Partnership, 2010).

Parallel to the debate around the semantics of 'participation' (Arnstein, 1969), the Pilots demonstrated multiple interpretations of 'co-design' and 'co-production'. The *C2* methodology's structure leant itself to enabling resident co-design and co-production, but examples occurred in varying states and stages throughout the action research cycle. As the final evaluation revealed, residents could be involved in co-production: as part of steering or project management groups; in co-leading the artistic activities; in participating and co-creating performance ideas in professionally-led productions; or in programming and delivering cultural and arts activities. Some resident groups led the process from start to finish. The final report six interpretations of co-design and co-production that took place across the Pilots (Merida, 2014: 31).

In the cultural co-design model, leadership roles have evolved into community curators or producers, co-ordinators, artist workshop leaders, event and project managers, community marketing leads with professionals working alongside residents sharing expertise. Volunteer roles expanded into outdoor event stewards, steering group members, catering, hosting and guiding roles. These roles were identified through the final Cultural Pilot evaluation (Merida, 2014), and whilst they might mirror the professional sectors of theatre, dance or visual arts, the distinction lies in the collaborative process between artistic practice and community engagement. As the producer noted:

> You have to love community engagement as much as your art form. It is important to work with artists who think that making a change in the local area is as important to them as their art. Some artists struggle to allow people to make their own structures and decisions and want to define their own art work to their own standards. (Evaluation Interview A1, 2013)

It became apparent that co-design and co-production provided a broad spectrum of possibilities for artistic engagement. Heartbreak Production's show would normally tour to a central location in Birmingham such as the Botanical Gardens, Edgbaston. However they were open to engage with new audiences and were persuaded to work with Hodge Hill Local Arts Forum to bring their production to Shard End. The children who participated were taught by a local youth theatre leader, employed by Reel Access for the first time to lead on this project. The feedback reveals both their passion for this area of work and their pride in the achievement of the young people engaged and the new audiences developed:

> Community engagement was very important for me – to see the young children involved have the opportunity to stand up on stage and be involved in such a high quality show and to see their faces when they 'owned the moment' of performance – was magic. Our research shows that most people came through recommendation or knowing someone in the show – only a handful came through seeing it on the internet or online. (Evaluation Interview A2, 2013)

To develop a co-design approach to commissioning work with under-represented groups requires a longer lead time. It worked in Balsall Heath where Ort Café wanted to work with the Somalian community. They worked to produce a 'Soo Dhawow' Somali Music Event with two local musicians in the café within OPW. The event co-ordinator at Ort Café commented that:

> We saw completely new people come through our door, inviting a whole new target audience into the venue so that we could start a conversation with this group about future joint-work. We also feel that we were able to empower residents to put on their own event and take responsibility for running sessions. This way we stepped back and were just able to host the events and the organisers were able to run projects as they wanted. We feel this was a success. (CCS, 2013a)

However in Shard End when working with DiVas, the initial request to co-commission local theatre organisation Women and Theatre was too ambitious. The partnership, brokered by Reel Access, was re-scoped into a process of drama workshops and role play to allow individuals within the group to understand the process of devising drama that could explore social issues. This process of co-design and devising the work around the nature of their own personal journeys was more relevant but they understood that it took time to gain the confidence and skills to convert personal experience into a performance. However, the opportunity to renegotiate the funding agreement with Reel Access and Women and Theatre and to experience how their concerns had been heard and that funding could be re-focused to suit their needs and aspirations was an important process in itself. DiVas will continue to work with Women and Theatre, and plan to apply for future project funding from the city council's Cultural Commissioning Service.

The most successful resident co-design and co-production came in Castle Vale with the creation of Castle Vale Festival. Since its regeneration, Castle Vale residents have been involved on various steering or management groups of: the Neighbourhood Partnership; the Community Regeneration Services; the Children's Centre; social care groups; the local Health Centre and Active Arts. As the Pilot co-ordinators, their initial report recommendations revealed:

> Residents and groups feel very strongly that the money should be invested in some big ideas that brings people together rather than dividing up into smaller monies that are isolated and have small impact … Residents want to celebrate the 21st anniversary of regeneration with a community event that brings groups together. (Active Arts, 2013b: 9)

The 20 strong steering group brought together many different groups drawn from Resident Associations, Community Regeneration Service (CVCRS), young people and staff from the local secondary school, a Sports group, the Neighbourhood Partnership Board, a local manager from Sainsbury's and the Chair and Co-ordinator of Active Arts who led the Cultural pilot. The ideas identified in the initial report were crystallised by this steering group into an action plan to create and deliver a Castle Vale Festival over 3 days. Creating a venue in a flamboyant Big Top set up on the flat playing field was a master stroke of planning that allowed everyone on the estate the ability to walk or take a bus to this location.

The opening ceremony involved 90 participants in a show that lasted well over an hour. It illustrated the journey of Castle Vale from the wartime airfield and runway where locally-built Spitfires were flown to other airports by women pilots for the first time. The images of flight were created as a dance by Year 10 performing arts students from Greenwood Academy co-choreographed by professional companies Highly Sprung Theatre and Vortex. A professional actor on a bicycle-powered 'time machine' narrated the story of resident engagement from living in high-rise housing to working with the Council to demolish and regenerate Castle Vale 21 years ago. The cast of local participants included children aged 7–16 from local primary and secondary schools and adults from the estate, most of whom had never danced or sung outdoors in a performance. This made for an emotionally charged experience that resonated with all who watched it (Active Arts, 2013a, 2013b).

A replica Spitfire loaned by the Spitfire Heritage Trust through the joint project with Leaps and Bounds and the secondary school completed the tangible links of arts with heritage. This event provided the Spitfire Heritage Trust with the opportunity to say a public thank you to Prince Seisso of Lesotho (received by his representative), as the first and smallest country of the Commonwealth Africa, to offer their financial support to build the Spitfires in 1940 (Spitfire Heritage Trust, 2013). Artistic co-production tasks included giving a voice to young people to programme the first 'Live and Loud' event and to invite the Luminites to perform. The Family Day involved stalls and fun fair events with football on the site led by local residents with bus tours led by local guides identifying key sites and stories from the Vale. The final evening involved the District's Arts Champions (Town Hall Symphony Hall) leading the 'Big Sing' choir event with the CBSO Berkeley Salon orchestra bringing the festival to a close. The event was a landmark for the area, with 90 participant performers, 38 volunteers to manage the event and over 5,000 audience members. The resonance of these events can be summed up by one steering group member: 'The Cultural pilot provided the opportunity to do something big, we seized the opportunity and it's made the case for us to do this again' (quoted in Merida, 2014: 22).

When collective effort is brought together to achieve a project greater than the sum of one person's ideas or actions there is the feeling of achievement that resonates across a community, as community arts practitioner Clements states:

> When rituals and ceremonies are being undertaken in indigenous cultures they are seen as 'healing' or as a 'remembrance' of how things should be. Most societies are so busy in their pursuit of 'individualism' they need a ritual or ceremony every now and again to bring its members back into the equilibrium. (Clements, 2004: 31)

As Durose et al. (2013: 331–332) note 'co-design offers a different approach to working with communities', involving them in identifying problems and through creative practice turning people into 'innovators and investors' rather than passive consumers. Two items on their list include 'Do be creative' and 'Do positively inspire people with what can be done' (ibid.: 331, 332) can be applied to these Cultural Pilots. The Castle Vale Festival, contributed to a larger social outcome with contributions to well-being.

The Role of the Artist and Co-ordinator in Co-design and Co-production

In some instances, the Pilot projects were artist-led, and in other instances artists were brought in to bring additional skills to residents, sometimes to raise creative aspirations or artistic quality to amplify the production. One of the artists working on the lantern parade with the Play Centre had honed her skills in working with international participatory outdoor theatre company Emergency Exit Arts. The excitement for the participants lay in creating something they could show in the dark and include their families in an evening parade.

The shadow puppet performance could only have developed once the trust with the children had been established. Together, the young people with the artists co-wrote a story about a flower 'Wishella' trying to grow up amongst the weeds and birds of the urban landscape, attempting to overcome the obstacles until she accepted herself for the beautiful flower that she was – not a weed! The sharing of this work in a packed family room with families of all diverse backgrounds and ethnicities provided a powerful message of self-acceptance and belonging within a community. The audience of families and children were very proud of their achievement and understood that creative activities could provide a sense of belonging to a local neighbourhood and venue (CCS, 2014c).

Some artists and companies came in to raise the bar and expectation of the level of arts activity that could be achieved. It could be said that the production in Kingfisher Park was an 'intervention' where an external arts organisation has taken an off-the-shelf product and applied it in Shard End. But the challenge for Shard End's Cultural Pilot was to create a local arts infrastructure, where residents could come to a local central place and experience and participate in culture (CCS, 2013b). There is always a danger that the 'parachuting-in' approach could cause more damage than good, however, in this instance, the partnerships and brokerage by the Pilot coordinator meant the event kick-started the appetite for using the park and turning what has previously been seen as a barrier into a piece of public land that connects communities. While the artists had been brought in from outside the area, there was a team of event volunteers who were local, who grew and took ownership of the event. Similarly, artists came in to Castle Vale for the creation of the opening ceremony of the Festival. Staff of the secondary school said that the young people working with the professional company to produce this performance 'really exceeded their expectations' (Merida, 2014: 25)

Equally the Castle Vale Pilot challenged pre-conceptions that arts activity is 'culturally elite' and too sophisticated. A CBSO ensemble played at the Festival and one steering group commended: 'we took a gamble with the Symphony Orchestra, introducing new arts and culture to Castle Vale, it turned out to be a sophisticated evening' (Merida, 2014: 26)

In this Festival and throughout the Cultural Pilots, co-ordinators played a key role in brokering new relationships with local artists, with Arts Champions and/or city centred arts venues. Without these trusted advocates the adventure to try out something new may not have been achieved. The Cultural Pilot evaluation report identifies 14 essential attributes for a successful co-ordinator, outlining a mix of community, arts and personal skills as well as events and project management experience (Merida, 2014: 69).

Building the Scaffolding: Developing the Local Arts and Cultural Infrastructure and Capacity for Local Residents to Engage

The third outcome of the Cultural Pilot refers to the creation of a stronger local arts and cultural infrastructure, and the final outcome is to develop the capacity for residents to engage in arts and culture.

In terms of building resident capacity, there was strong involvement from volunteers and participants across the three Pilots, which extended the reach, depth of engagement and skills development with local people. The need for this varied between areas and events, for example volunteers for Balsall Heath Carnival were recruited through the local church and needed very little training, whereas in Shard End volunteers were formally trained over a two week period and gained an SIA Event Security qualification. Other producers or artists were given informal shadowing opportunities (the youth theatre leader in Shard End continued to lead on the second year of theatre in the park during 2014).

With reference to both these outcomes, informal partnerships with local venues such as libraries and parks have been sustained into a second year of delivery and staff communication has been strengthened for future shared cultural events. The steering group in Castle Vale has continued to plan events including a second Festival. The Cole Valley steering group has emerged from Shard End Ward and they have planned the annual arts calendar for the ward. In Balsall Heath the partnership between St. Paul's Community Development Trust, OPW and the local Play Centre has been strengthened and a new Hall Green Local Arts Forum was constituted as an unincorporated association during 2014 with those venues leading the way. The Yorkswood Resident Association used the evidence from the Griffins day to submit a successful Heritage Lottery Fund application to complete the restoration of the remaining five stone Griffins on the estate as a direct result of the work undertaken through the Cultural Pilot.

The Pilots provided opportunities for people to not only activate or reactivate an interest in engaging in the arts, but also in other local community initiatives. For example, one interviewee from Shard End linked the resulting raised sense of local awareness with the large turnout at a public meeting to discuss a new leisure centre, which took place after the production (Merida, 2014: 42).

Building audiences to engage in these events involved developing community marketing. The Balsall Heath Carnival was enhanced by the 200 butterflies hung along the route of the procession and the poster design decorated by children. In Castle Vale, the Knits and Pieces craft group came up with the 'Castle Valiens' a marketing campaign with a difference: imagining extra-terrestrials had landed on the estate. Knitted creatures by the local arts collective popped up all over the Vale on public art, in Sainsbury's supermarket, on street corners as promotion in the lead up to Castle Vale Festival. The Facebook and Twitter following made it to the front of local newspaper the Tyburn Mail and captured the imagination of the local population, who spotted the Valiens in the supermarket or hung on lampposts.

The strengthening of local infrastructure through this process was described by one co-ordinator as like 'building the scaffolding to enable people to grow and develop by providing information and guidance in a timely manner, not rushing or overloading people but facilitating and supporting their decision-making so that people become confident in their own decisions' (Merida, 2014: 33). Sometimes policy makers can confuse the hard infrastructure development as buildings or transport, with the equally important 'soft' cultural infrastructure which builds social capital and effective social networks. The success of the Cultural Pilots laid in the latter and will be proven over a period of years not months.

Conclusion

This chapter has provided an overview of the context, methodology and operation of the Birmingham Cultural Pilot programme. A critical analysis of the three Pilots has been undertaken regarding the extent to which they achieved the agreed outcomes, and in particular, the success of cultural co-design and co-production in building on existing infrastructure, bridging gaps and localising neighbourhood provision.

While both Castle Vale and Balsall Heath had long-term experience of social and community engagement resultant from former neighbourhood engagement programmes and development work, it was evident that cultural co-design and co-production was ambitious. As pointed out by Merida (2014: 27):

> It was quite an ambitious desired outcome for neighbourhood-level groups to be ready and able to co-design and co-produce arts and cultural activity […] particularly in areas like Shard End where there was no cultural foundation to build upon, but also in Castle Vale where it required a step-change from (relatively) passive to active engagement on the part of local people.

Despite this, all Cultural Pilots successfully lifted many of the main barriers to Birmingham residents participating in culture, as previously identified by Ecotec (2010). The Pilots provided: free arts activities to break down the barrier of unaffordability; raised awareness of local amenities and attractions through partnership collaboration; in some cases challenged perceived cultural elitism; and enabled people to walk to a local community venue to access culture.

Recommendations from the Cultural Pilots evaluation focused on the continued adoption of the *Positively Local C2* methodology involving local co-ordinators working with local community groups. Lessons learnt highlighted the need for strong local partnership development to support the co-production and co-designing of arts and cultural activities. This requires local leadership and strong communication based on mutual trust and dialogue to create an appropriate range of cultural activities in line with the needs of residents and priorities for each area. Sustainability of these pilot activities has happened in Castle Vale and Shard End with a continuation of The Castle Vale Festival 2014 and Heartbreak Production in Shard End 2014. In Balsall Heath the various independent programme of arts activities have continued and a new constituted organisation Art Works Balsall Heath has been set up to steer development. Residents are continuing to be involved in steering groups, planning and delivering local events that communities want to engage in.

All Cultural Pilots performed against the original outcomes for the programme, leaving a legacy of increased local resident participation in arts and cultural activity and experience in resident co-design and co-production of local arts and cultural activity. This has created an appetite for creating events with other local partnerships. This has begun to extend the capacity of residents to create and participate in new activities in local neighbourhoods, try out new and different art forms and become audiences to events in non-traditional spaces.

Feedback from participants, co-designers, co-producers and coordinators reaffirm the statement within BCC's Cultural Strategy, that 'participation in culture is inherently a good thing' (Birmingham Cultural Partnership, 2010: 4). However each Pilot has shown how co-production and co-design involves residents in expanding their perceptions, being challenged to take on new roles in terms of planning and delivery of events, creating a sustainable way forward. Where participation in events has created moments of personal reflection, the experience of the arts has enabled people to extend their skills and understanding and contribute to the world they live in. These Pilots have also shown how culture can contribute to other cross-cutting agendas, such as the Community Based Budgeting pilot priorities of health and wellbeing, civic engagement, economic impact, development of transferable skills and improved environment. They have engaged with unemployed residents and communities where arts participation statistics have previously been low. Using the *C2* methodology, the Pilots have brought about a step-change in localised community arts engagement, bridging gaps between services and creating relevant local provision on the doorsteps of residents. In some instances, this has led to greater involvement in other public agendas, such as public health and safety. There was a larger turnout to a public meeting in Shard End, which could be linked back to the Cultural Pilot event (Merida, 2014: 42).

Through the use of the *C2* model, listening to communities and learning what the issues are from residents first-hand, the Cultural Pilot process has developed trust, as participants in the DiVa project report (CCS, 2013b) commented: 'If anything is happening in the area that you or your [partner] companies are doing then we'd be encouraged to come because we've built that relationship'. The Pilots have grown people in each of the areas, sowing the seeds and nurturing the shoots for further community co-designed cultural intervention work in the future. The Pilot evaluation conveyed that people feel they have been listened to about arts and culture in their area, and people felt their opinions were valued.

Both the new *Social Inclusion Strategy* (BCC, 2013b) and Neighbourhood Strategy *Transforming Place* (BCC, 2014d) were adopted by BCC towards the end of the Cultural Pilots. The latter makes reference to the role of the Local Arts Fora and Arts Champions in 'developing a place-based approach to engaging communities in the heritage of Birmingham and using this as a tool for social regeneration' (BCC, 2014d: 27), showing an increasing recognition for the role arts and culture plays within society and shared outcomes. As a result of the Cultural Pilots, additional money from BCC, DCLG and ACE was secured to roll out the pilot concept to all 10 districts in 2014–15.

The 'Connecting Communities through Culture' programme will continue to test out and embed a different way of allowing residents and communities to create art and culture – allowing and enabling different voices to be seen and heard. Technology is providing exciting ways in which these voices can be shared with new tools of engagement that are re-defining and blurring the boundaries of 'cultural elitism'. It is evident from these culture pilots that cultural co-design strengthens participation and builds skill in community engagement and event management, which are transferable skills. It is hoped that the value of cultural engagement and positive experiences evidenced in this work will continue to be taken seriously by political and cultural policy makers as vital contributions to art and culture, and to social inclusion, public health and well-being. As former Prime Minister John Major (2014) argues, the arts go beyond political party divides:

> The arts are not an add-on to people's lives, not an optional extra. Art is integral to life. It enhances it. It civilises and helps build rounded personalities. It encourages people from other countries to visit us and promotes cultural understanding, which in today's world is needed more than ever.

References

Active Arts (2013a) *Castle Vale Cultural Pilot: Castle Vale Festival Report*. unpublished.

Active Arts (2013b) *Final Report: Community Based Budgeting Cultural Pilot Castle Vale*. unpublished.

Active Arts (2014) *Social Prescription Project: Arts on Prescription Pilot*. unpublished.

Allen, F. and Dickson, M. eds (1995) *Art with People*. Sunderland: a-n Publications.

Arnstein, S. (1969) A ladder of participation. *Journal of American Institute of Planners* 35(4), 216–224.

BCC (2009) *A Creative Future II: A Strategy for Children, Young People & the Arts* unpublished.

BCC (2012a) *Leader's Policy Statement* http://www.bvsc.org/sites/default/files/files/LeaderPolicyStatement_FINAL.pdf accessed 22 August 2014.

BCC (2012b) *Community (Neighbourhood) Based Budgeting Culture Pilots Project Programme*. unpublished.

BCC (2013a) *Arts Champions Scheme* http://birmingham.gov.uk/artschampions accessed 22 August 2014.

BCC (2013b) *Making Birmingham an Inclusive City White Paper* http://www.birmingham.ac.uk/Documents/college-social-sciences/public-service-academy/white-paper-march-2013.pdf accessed 22 August 2014.

BCC (2014a) *Population and Census* http://www.birmingham.gov.uk/cs/Satellite?c=Page&childpagename=Planning-and-Regeneration%2FPageLayout&cid=1223096353755&pagename=BCC%2FCommon%2FWrapper%2FWrapper accessed 22 August 2014.

BCC (2014b) *District Strategic Assessments* http://fairbrum.wordpress.com/about/district-strategic-assessments accessed 22 August 2014.

BCC (2014c) *Cultivating Culture: Celebrating Local Arts in Birmingham*. unpublished.

BCC (2014d) *Transforming Place: Working Together for Better Neighbourhoods* http://fairbrum.files.wordpress.com/2014/04/transforming-place-working-together-for-better-neighbourhoods-framework.pdf accessed 22 August 2014.

Birmingham Cultural Partnership (2010) *Big City Culture 2010–15 Birmingham's Cultural Strategy* http://birminghamculture.org/files/bhamculture_report.pdf accessed 22 August 2014.

CCS (2013a) *Evaluation Report Jo Reichert, Ort Café*. unpublished.

CCS (2013b) *Final Co-ordinator Report for Balsall Heath Cultural Development Programme, Midan*. unpublished.

CCS (2013c) *Shard End Cultural Pilot Report, Reel Access*. unpublished.

CCS (2014a) *Evaluation Report St Pauls Community Development Trust*. unpublished.

CCS (2014b) *Evaluation Report Some Cities*. unpublished.

CCS (2014c) *Evaluation Report Sophie Handy Old Print Works*. unpublished.

Clements, N. (2004) *Creative Collaborations*. Swansea: Sound of the Heart.

DCLG (2011) *A Plain English Guide to the Localism Act*. London: DCLG.

DCLG (2013) *Neighbourhood Community Budget Pilot Programme; Research, Learning, Evaluation and Lessons* https://www.gov.uk/government/uploads/system/uploads/attachment_data/file/224259/Neighbourhood_Community_Budget_Pilot_Programme.pdf accessed 22 August 2014.

Durose, C. Richardson, L., Dickinson, H. and Williams, I. (2013) Dos and don'ts for involving citizens in the design and delivery of health. *Journal of Integrated Care* 21(6), 326–335.

ECOTEC (2010) *Identifying the Barriers to Cultural Participation and the Needs of Residents of Birmingham* http://birminghamculture.org/files/Untitled-Folder/BarrierstoCulturalParticipationinBirmingham22072010amended.pdf accessed 22 August 2014.

Garry, K. and Goodwin, P. (2014) *Igniting a Spark: An Evaluation of the Birmingham Cultural Pilots Programme*. unpublished.

Gillespie, J. and Hughes, S. (2011) *Positively Local: C2 a Model for Community Change*. Centre for Welfare Reform Policy Paper, University of Birmingham, Birmingham

Local Government Association and DCLG (2013) *Our Place* http://mycommunityrights.org.uk/wp-content/uploads/2013/06/Our-Place-and-what-the-pilot-areas-achieved.pdf accessed 22 August 2014.

Major, J. (2014) The arts are not an 'add-on'. *Create: A Journal of Perspectives on the Value of Art and Culture* http://www.artscouncil.org.uk/what-we-do/value-arts-and-culture/state-arts/create/culture accessed 15 August 2014.

Matarasso, F. (1997) *Use or ornament? The Social Impact of Participation in the Arts*. Stroud: Comedia.

Spitfire Heritage Trust (2013) http://www.spitfireheritagetrust.com/#!tribute-programme/c14gj accessed 22 August 2014.

Vector Research (2011) *Final Report P1473 Cultural Participation* http://birminghamculture.org/files/Untitled-Folder/p1473BCPCulturalneedsreportFINAL.pdf accessed 22 August 2014.

Webster, M. and Buglass, G. eds. (2005) *Finding Voices Making Choices: Creativity for Social Change*. Nottingham: Educational Heretics Press.

Chapter 7
The Everyday Realities of Digital Provision and Practice for Rural Creative Economies

Liz Roberts and Leanne Townsend

Introduction

Conceptions of 'the rural' are shifting to encompass a more relational understanding of rurality as a 'multi-authored and multi-faceted space, constituted through local-global interconnections and their place specific, sometimes contested, manifestations' (Heley and Jones 2012: 208). Rurality is increasingly framed through concepts of networks, connections, flows and mobility, and recent research explores the scalar interactions that comprise rural places, and the 'diverse networks and flows that criss-cross rural and urban space' (Hoggart, 1990: 43). Rural places retain a politics of location and differentiation as global connectivities are played out and negotiated through relative *permanences* in people's everyday lives (Heley and Jones, 2012). Work on the rural creative economy figures this tension in terms of place distinctiveness and scales of interaction. For example, developments in internet technologies mean that rural creative practitioners have access to expanded, global markets, however, often their work can be inspired by and meaningful to consumers because of their rural location, for example as part of a tourist experience (Craft Council, 2011). The continual shaping and reshaping of the rural means that 'careful examination of the situated and practiced connections between the global and the local' is required (Heley and Jones 2012: 214). We explore the role of digital practices in the rural creative economy within this relational understanding of the 'global countryside' (Woods, 2007). Whilst much work on creative economies takes a place-based approach, often looking at a specific city, region or hub as a case study, this chapter focuses on individuals living in remote rural locations across Scotland who are engaged in different creative practices. What they have in common is that their choice of rural location means that they typically have much lower internet connectivity than the UK average. Rather than think in terms of sectors or networks, we explore the everyday practices of rural creatives. Whilst some creative economy studies note the limitations of rural internet connectivity, few have examined the realities of ICT for rural creatives: their everyday uses, desires, frustrations, or how they adapt and negotiate within competitive markets and increasingly digitised transactions. The rural creative economy literature on the whole refers in passing to the opportunities of internet technologies to have greater 'reach' (Herslund,

2012). This chapter adds to existing understandings by drawing lessons from in-depth interviews carried out with practitioners in remote rural Scotland.

We explore the everyday realities of digital provision and practices for creative practitioners living in remote rural locales. The chapter outlines the diverse realities of digital practice and the distinct politics of location that are played out through and against these: the creative ways they navigate the professional challenges of remote rural living, particularly the lack of digital connectivity. This is set against the UK policy discourse of the rural digital economy as one of 'potential'. Recent research on rural creative economies suggests that it exists through regional networks and hubs, as well as city-facing practices, and that it takes place predominantly in accessible rural places. In response to this, authors look to the distinctiveness of rural locations with different practices and the rural landscape as a source of inspiration (Gibson et al., 2010; Ward and O'Regan, 2014). This chapter explores the complex, multi-scalar activities of creative workers, thereby contributing to reducing the knowledge gap between rural and urban areas and work that looks at rural creativity in 'the margins' (Gibson, 2010). Creative workers are expected to contribute to rural resilience through bringing diversification to local economies, as well as skills, resources and employment opportunities. In our literature review, we examine the opportunities of the rural creative economy within rural, digital and creative economy policies in the UK. Then, through consideration of the findings of our research with rural creative practitioners, we look at the current situation, to better understand the everyday practicalities of digital aspects of creative work in rural contexts.

The Rural Creative Economy

The rural creative economy has received considerable attention recently in both policy and academic literature. In the UK, the cultural industries sector accounts for 5.6 per cent of total jobs (DCMS, 2015). In both employment and gross value added (GVA), it has grown faster, on average, than the UK economy as a whole in the last five years. Creative professionals contribute £500m to the UK rural economy every year, supporting tourism and bringing 'new ideas and ways of working to rural areas, as well as new products and services, attracting other businesses and investment' (Craft Council, 2011). The UK creative economy is defined as encompassing advertising, architecture, arts and antique markets, crafts, design, designer fashion, film, video and photography, software, computer games and electronic publishing, music and the visual and performing arts, publishing, television and radio (DCMS, 2001). Rural creative economy research, however, blurs these categories, suggesting that rural creativity takes place formally and informally (Thomas et al., 2013), consisting of sole traders, community development partnerships, and private companies (Gibson, 2010) and that creative practitioners switch to different types of practice throughout their career or work in multiple sub-sectors simultaneously (Herslund, 2012). Studies of the rural creative economy are varied, for example, focusing on creative in-migration (Wojan et al., 2007), policy and practice (Bell and Jayne, 2010), individual sectors such as crafts (Harvey et al., 2012; European Commission, 2009), and a broader conception of knowledge-based sectors (Skoglund and Jonsson, 2012).

Calls have been made to address the 'distinctive arts ecology' of rural areas, for example, the different working practices, markets, aesthetics and role within communities (Bell and Jayne, 2010; Gibson, 2010). Rural creativity research speaks to the precariousness of the sector and suggests a large number of rural creatives are self-employed micro-businesses (Bain and McClean, 2012; Donald et al., 2013; Herslund, 2012). Central across this work is the conviction that a 'one-size-fits-all' model for the regional development is not appropriate to rural contexts. Rural creative practitioners are working in different environments to the 'buzz' or creative quarters of Florida's (2012) creative cities. But this environment is no less creative. Remoteness has been equated with: 'limited types of creative making; wariness of newcomers and new ideas; the loss of young people; limited access to business expertise, production services and training; lack of cultural stimulation; and high transport costs' (Gibson, 2010). Rural arts and crafts can also fall victim to 'cultural cringe' when it is considered parochial in comparison to what is produced in urban centres (Bell and Jane, 2010). Yet growing attention from policy-makers in rural areas has made the rural creative economy an important focus of research. In the sections that follow we develop this attention by considering the interconnected factors of digital practices and policy for rural creative work.

The Digital Promise

Rural resilience, alongside refocused attention on the post-productivist 'local' and 'small-scale', and the desire for a 'living countryside' with 'dynamic local actors' (McDonagh, 2013), is a growing area of concern. Rural creative industries, within the broader context of the digital economy, is considered a key approach to diversification of the rural economy (European Commission, 2005). In the last ten years or so creative industries and digital policy have been brought together to implement rural development strategies. Europe's Digital Agenda and 2020 strategy is to provide growth and jobs in a sustainable and inclusive manner (European Commission, 2010a). European Union Rural Development recognises the need for diversification of the rural economy and growth of non-agricultural sectors, such as tourism, culture and heritage, and information-based service sectors (European Commission, 2005) and the opportunities of ICT for rural businesses (European Commission, 2010a). But McDonagh (2013) notes that whilst policymakers place high expectations on the rural non-farm economy, they seem less aware of ways to support and stimulate it in a robust way.

A key part of the UK digital agenda is to stimulate rural markets increasing demand and competition for digital infrastructure. *Britain's Superfast Broadband*

Future (DCMS, 2010) set the context for this with the goal to provide broadband of at least 30mbps to 95 per cent of the population by 2017, predominantly through new optic fibre infrastructure, via a market-led approach. A large part of the remaining percentage of the UK to receive connectivity will be rural areas, which lack the economic incentive for large telecommunications companies to 'roll out' the necessary infrastructure. Broadband Delivery UK (BDUK), a £530 million programme for rural broadband initiatives, was set up to respond to this shortfall.

The UK approach primarily requires remote rural communities to create their own access with the help of a Rural Community Broadband Fund (RCBF), for which they must bid (DCMS, 2010). Funding has been limited through RCBF; the scheme has been somewhat controversial, and its future is unclear (Guardian Online, 2013; House of Commons, 2013). Especially important for the rural creative economy is broadband infrastructure that is future-proofed and able to accommodate cutting-edge technologies. This is far from the reality for most rural communities who are dependent on British Telecom, which holds the monopoly on current UK broadband and phone-line infrastructure and policy commitments to a minimum of 2mbps. This urban-rural digital divide is illustrated by current average bandwidth speeds in the UK, which contrast starkly across urban (33.4 mbps), suburban (22.9 mbps) and rural (13.6 mbps) areas (Ofcom, 2014). Within Scotland, where our interviews were conducted, 'Increasing rural broadband' was placed at the top of priorities for rural areas (Scottish Government, 2011a). Like the wider UK policy, a bottom-up approach to broadband provision was identified as necessary in rural areas. Scotland's Community Broadband strategy has recently been injected with £2.5 million with the government restating its commitment to helping rural communities establish their own networks, and bringing total funds up to £7.5 million alongside £410 million to ensure that 85 per cent of Scottish people have fibre access by the end of 2015 (Community Broadband Scotland, 2014). There is also recognition that those in the remotest areas cannot expect a guarantee to access (DEFRA, 2013), which is the case for many of the practitioners we have worked with.

Even so, rural and digital policy posits broadband as a driver of growth with challenges to realising its potential in rural areas. The UK identifies economic benefits of broadband access and services as: driving innovation, profitability, R&D and trade; boosting jobs; enabling productivity and diversification, such as creative industries growth (Welsh Assembly, 2010; Scottish Government, 2011c). The UK government acknowledges that business gains from ICT will be dependent on skills, management and willingness to invest and that economic spillover may be uneven within rural regions (European Commission, 2009). For example, the Commission for Rural Communities (2009: 20–22) argues that variable broadband availability and speeds within and between rural communities has the potential to divide rural communities into connected and unconnected areas, leading to negative consequences for rural businesses. Scotland's rural economic policy seeks to build competitive advantage and make rural areas attractive places in which to do business through raising awareness of e-commerce and creating initiatives and partnerships to increase Scottish business internet use and ICT skills (Scottish Government, 2011a; 2011b). A number of initiatives such as the 'Getting British Business Online', the 'Scottish Business Portal Programme' and 'Interactive Scotland' are partnerships to increase Scottish businesses' internet use, awareness, ICT skills and digital participation (Scottish Government, 2011a). Policies are oriented towards helping rural businesses to adapt and upskill. Rural businesses reportedly, however, still lag in their use of internet. For instance 25 per cent of Scottish businesses are not using internet (Scottish Government, 2011b; Galloway et al., 2011), perhaps reflecting that access to internet and ICTs does not equate simply with their use (Salman, 2012) but also includes a number of cultural, social and attitudinal factors (Helsper, 2012).

Interlinking Creative and Digital Strategies

At European Union and UK level, creative and digital economies are closely linked. Cultural and Creative Industries (CCIs) and their role in the new 'experience economy' are considered key to the 2020 strategy for growth (European Commission, 2010b). The European Union states it is committed to creating 'appropriate and favourable conditions' to enable CCIs to integrate new digital technologies (European Commission, 2010b). Within the UK, the *Creative Britain* report (DCMS, 2008) stated that creative industries must rise to the challenge offered by digital technology, by harnessing the strategic ability to understand and maximise its impact. Reciprocally, 'cultural contents' are also viewed as playing 'a crucial role in the deployment of the information society, fuelling investments in broadband infrastructure and services' and digital technologies, contributing to development of e-skills and media literacy (European Commission, 2010b: 3). The need for constant reskilling and upskilling of the creative workforce in light of rapidly changing technology is noted as a barrier to sector growth (Northern Ireland Assembly, 2013). CCIs, like other rural or micro-enterprise businesses, face the challenges of costs, skills development, constant technological development and knowledge of digital rights management when 'going digital' (European Commission, 2010b). New business models are recognized as necessary for development of the cultural sector in rural regions, enabling innovation and new technologies (European Commission, 2010b).

Scottish digital policy identifies that digital technologies are a significant driver of innovation within creative industries, that they create new markets, and have 'spillover', contributing to growth in other areas. It stresses that the public sector must offer a supportive approach to help businesses seize digital opportunities and that creative industries can be transformed by digital innovation (Scottish Government, 2011c). Scotland's Creative Industries Strategy outlines how digital media companies can access market research, product and technology support through various initiatives. A particular policy goal is to become more joined-up, linking to the Technology Strategy Board and NESTA (National Endowment for

Science, Technology and the Arts) (Scottish Government, 2011c). On the whole, Scottish creative practitioners would appear to be well supported and their role within the rural economy acknowledged through these various structures.

Broadband internet access is increasingly recognised as a key enabler to creative business success in rural areas (Bowles, 2008; Burns and Kirkpatrick, 2008). As creative and digital practices become interlinked (Hartley, 2005), the internet performs an increasingly important function as a marketing tool enabling creative workers to connect to others, to sources of inspiration and trends, and to promote themselves in global and alternative markets (Duxbury and Campbell, 2011). Duxbury and Campbell (ibid.: 114) find that the broader context for rural cultural activities is 'the availability and capabilities of broadband internet, which is particularly highlighted by rural communities' growing desire to attract the 'creative class' as residents-with-businesses. Further, they argue that 'the interconnected world of creative production is more complicated than the image of a simple "city-country divide" and should focus on networks and flows of people, information, and creative production' (ibid., 117). In contrast, Galloway et al. (2011: 254) writing about rural businesses in general claim that 'the importance of the local when applied to rural small firms' internet use is largely neglected'. Yet their finding that rural firms are interested in local markets and are not necessarily driven by or even desirous of growth may now, due to the speed and diffusion of technology development, already be out-dated. Bell and Jayne (2010: 216) note that the use of new technologies 'raises an important symbolic issue that highlights disparities between policy makers' conceptions and practitioners needs'. This chapter examines this disparity through looking at the realities of digital practices for rural creative practitioners living in remote places. It explores the extent to which practitioners can 'rise to the challenge' offered by digital technology given their low connectivity.

Rural Creative Practitioners and Internet Connectivity

In this section we reflect on the findings from fifteen in-depth interviews with rural creative practitioners working in remote rural regions across Scotland during 2013. The research was carried out in the context of a large project investigating the potential of satellite internet connectivity for remote rural creative workers who could not access cable broadband. Much work on the creative economy focuses on a place-based case study, such as a city or a rural region. Instead this chapter draws together the experiences of those who have a remote rural location in common; however, these locations are neither necessarily proximate nor alike. As 18.1 per cent of the Scottish population are estimated to live in rural areas (with 11.6 per cent in accessible rural areas, 3.4 per cent in remote rural areas and 3.1 per cent in very remote rural areas, Scottish Government, 2012), they represent the reality of digital connectivity for a significant proportion of the Scottish population.

The creative practitioners interviewed reflect a mix of sectors and work approaches. Whilst many were in-migrants, many had lived in rural locations for 20 or more years. The majority had had a career change and many worked in more than one area of creative practice, such as the lighting designer who was also a novelist, and the music producer who made films. Distance was not always the biggest factor in creating this. Often remoteness was signalled by sparse population density in an area, the lack of transport infrastructure or type of terrain e.g. mountainous/forested (particularly the case for our participants based on the smaller, more remote Western isles of Scotland). Respondents reported that lack of internet did contribute to their feeling of remoteness, and whilst some felt that the rural landscape was important for their creative practice, on the main part the decision to live in these kinds of locations was either for family reasons, because it was where they had either been brought up or always lived, or to pursue a more laid back lifestyle away from the 'rat race'. For some it was not a choice but necessitated by personal factors.

A proportion of our respondents fall into the category of what Andrew and Spoer (2011) call 'digital creatives' using digital technologies as part of their creative practice, such as video, editing suites, and production software. Whilst none of the respondents fell into a category commonly discussed in rural creative economy literature – rural crafts – a group of respondents were artists who also worked within arts education and curation. Another category incorporated in our sample is that of rural arts hub or arts organisation. Other respondents represent writing, photography, film production, radio production, marketing and design. Some, but not all of our respondents worked and lived in rural areas considered in economic decline. Other predominant industries in the areas in which our respondents lived include tourism, food and drink, and agriculture. Our respondents represent a range of remote rural settings, on both islands and the Scottish mainland.

General Technology Use

Perhaps one of the first things to note is the amount of technology the rural creatives are working with. Many interviewees were working across one or more laptop and several PCs or smart phones:

> So we juggle satellite technology, mobile phone internet access, local internet access and we quite often steal other people's using BT (phone) and things like that. (Radio producer, Aberdeenshire, 2013)

> We use it all the time and we've got seven computers and a laptop … . we do an awful lot of work on Excel, we're a research company so we do lots of collation, charts and graphs and pie charts. We run – I wouldn't like to say how many databases. (Marketing business, Aberdeenshire, 2013)

Some were using internet-enabled technologies as part of their creative practices and others used it for the business or marketing side of their practice. The potentials and realities of internet-enabled technology use were different for each of them.

In their general use, each of the creatives had a level of web visibility and were active online in multiple ways. For many of the respondents, email was the prime form of communication being used to contact clients and colleagues. The internet speed that respondents were receiving varied from half a megabit to two megabits per second. For most, email only became a problem when it involved attaching large files, which often was a daily and essential part of their business. Several respondents had to accommodate travel into their daily routine in order to access email.

> Email mainly. Because you know it can work and you just get totally hooked into it until you become dependent on it. (Radio producer, Aberdeenshire, 2013)

> I make sure I check my emails at least once a day but that does involve me driving six miles down the track. (Lighting designer, Aberdeenshire, 2013)

In order to function digitally, our respondents engaged in a number of creative practices to work around their lack of connectivity, such as uploading files overnight, going to the library, homes of friends, or the pub car park for wifi access, or doing certain tasks at certain times of day:

> It depends what we're doing because for a fast turnaround we do it on a laptop with a dongle to connect you if and when you can get a reception. But mostly with the bigger stuff I'd be coming home, editing it at home and sending it in. (Radio producer, Aberdeenshire, 2013)

> I'd be sending large files to people, whereas at the moment I copy them onto CD, stick it in a jiffy bag and take it on one of the three buses to Ellon [post office]. (Photographer and creative writer, Aberdeenshire, 2013)

Although these responses to poor connectivity evidence adaptability, our findings suggest that the need to adapt caused our respondents frustration, isolation, anger, boredom and anxiety:

> I can get about 2 megabits. If I'm sending a programme back it will take hours, it will get there but sometimes it will crash on its way … it gets your heart going. If you try and send something back on a tight deadline then you really get quite nervous. It doesn't always work … (Radio Producer, Aberdeenshire, 2013)

> It's very isolating, its stops you competing at the same technological level … it's one of the biggest struggles I have with everything. And it's a bore that it's a struggle. (Children's video producer, Aberdeenshire, 2013)

One respondent notes that it takes them half a day to do something that would take five minutes to do for others in better-connected locations. Rural creative practitioners therefore appear to spend undue time attending to basic administrative tasks, sapping energy that would otherwise be put into more creative tasks. It also illustrates both the level of commitment that rural creatives are giving to their professional activities and their desire to live in remote locations.

Respondents were keen to take on the 'cost' of remoteness themselves and stated that they often withheld their lack of connectivity from clients until they had secured a contract, due to fears of appearing unprofessional. Some of the practitioners had been in a position where their low connectivity had lost them business:

> People like you to be in the city, where they can physically see you. If they can't trust the technology is going to work, you don't get the gig. And I've been in that situation because of dodgy internet connections. People made you drive to Glasgow to record a voice track even though you could do it at home, they just don't trust it. (Radio producer, Aberdeenshire, 2013)

> I try to make it completely invisible that I live here. If you haven't even got a fast internet connection, it's an association people would make. It's like, 'if you can't even send me a file quickly, do they really know what they are doing?' (music and film producer, Western Isles, 2013)

Inadequate internet connectivity also impacted on their ability to innovate because it meant that they were less able to keep up with developments in their field and with what competitors were doing online. Whilst it seems fairly low tech, this was a key point mentioned by respondents, also relating to their feelings of isolation and to their capacity for professional development:

> Because I don't spend any or much time online, I am not really that up to date with technical developments in my field. I know about them but I don't have a means of researching them. (Lighting designer, Aberdeenshire, 2013)

> It just would be nicer to have better connectivity with people and know what they are doing as well because we all inspire each other, don't we? (Artist and arts educator, Aberdeenshire, 2013)

The respondents on the whole had good awareness of the digital resources available to them and were often limited only by their lack of access. Pragmatic solutions were sought such as visiting a library but these were not always appropriate methods when issues such as security for financial transactions and intellectual property were concerned:

> I have a close friend down in Edinburgh and I go there because certain use of the internet will entail looking at my bank statements and shifting funds around and things like that and I would never do that in a library situation … I've got to be quite careful that people don't steal my images. (Artist, Tayside, 2013)

Indeed, the respondents were well aware of how they would exploit better connectivity in their business activities, with many having a 'wish list' of applications that they would engage with:

> You can work round it, you can make life easier and better and faster and with a faster broadband what it would allow us to do, my colleague lives in a rural area as well, with a decent speed of broadband we could do the whole programme live, we wouldn't have to go to Aberdeen to do it, I could do it from home, he could do it from his home. (Radio producer, Aberdeenshire, 2013)

> I want to be able to use Skype reliably any time. I want to be able to be in a position where I can go to, at some stage, when the technology comes, that we can do video conferencing and things so you aren't always driving and travelling and it's easy, it's not difficult. I want to be able to see YouTube whenever I want. I want it to be fast. I want us to be able to use it over the house so that it's easy and I want to be able to use it on my television set so we just function like anyone in the rest of the country a bit. (Children's video producer, Aberdeenshire, 2013)

The general usage of internet and digital technologies reflects a mix of local and low-tech responses as well as significant time investments from rural creative practitioners. Their capacity to network, collaborate and promote themselves, reflects similar negotiations and compromises, but foremost reflects that rural creative practitioners, on- and off- line are nevertheless embedded in global systems of production and consumption as the next section outlines.

Networking and Collaboration

In investigating the day-to-day digital practices of rural creative practitioners it became clear that they could not be defined as strictly 'rural' in any sense. All were engaging face-to-face and virtually with colleagues, clients and customers at a regional, national and international level. A combination of face-to-face and online interaction was viewed as the best way to gain and maintain relationships:

> Because of the nature of my business it would be face-to-face meetings, but in between that obviously it's emails. All the time. (Marketing business, Aberdeenshire, 2013)

> I find my work in three ways: way one is that people find me … through word of mouth. I'm a member of the Association of Lighting Designers so I've had work from people looking through the directory of members and looking for specific skill sets … there's looking at the trade press and responding to ads on that and there's things like Scotnits [an email group] and the Creative Scotland website. (Lighting designer, Aberdeenshire, 2013)

It was the combination of online and offline connections, promotional and networking activities that kept rural creative practitioners in business.

Many respondents commented that in terms of professional networking and collaborating, better access to 'real time' communication technologies (such as Skype and Facetime) was desirable.

> We also work with a lot of people, as far as Australia, who do some of our designs and all that sort of thing. My partner is now based in London … I need to talk to her regularly and we have issues around that … I need to be in a situation where I say 'we're having a meeting on Skype' or if I'm bringing any other people in, it needs to work. It's not a faffing business – I'm trying to actually make this work and if you want to do it professionally you have to do it at that level. (Children's video producer, Aberdeenshire, 2013)

> Skype is really important … an artist just had an interview with *Sculpture* magazine – trying to conduct it over Skype was pretty horrific. (Curator within arts development organisation, Aberdeenshire, 2013)

> I would do more if I could communicate if I was able to Skype … I am missing out on the communication side of things quite badly. (Lighting designer, Aberdeenshire, 2013)

Although based in rural, often remote locations, some of our respondents felt that in order to survive economically, they needed to collaborate with and work for remote colleagues and clients. Generally they felt that, given adequate access to technologies, they would be able to run their businesses from any location. Connecting via technology goes some way towards reducing the distance barrier, but needs to be combined with actual travel to other locations – many were working for periods of time in different locations both across the UK and abroad:

> If I was to sit here in isolated [village] then no, I would not get business … I've been to Glasgow, Perth, I've worked in San Francisco, I've worked in Houston – if you want something, it doesn't really matter where you live but you do need the technology … without computers, without things like that we couldn't possibly work here. (Marketing business, Aberdeenshire, 2013)

> Well, in terms of the work I do internationally it really matters nothing where I live … for the last six years I've been working for a company that's based in Sheffield and as far as they have been concerned it didn't really matter where you lived; they are an international company. (Lighting designer)

Others were working mostly with local collaborators and clients, and were less likely to travel widely themselves:

> It's very much in the local area up here, the DVD has been using local children … and it's really marketed at the moment on Amazon and through selected stores. (Children's video producer, Aberdeenshire, 2013)

> I find the rural environment more suited to my style although I feel I do keep in touch with what is going on in the cities. (Artist and bagpiper, Tayside, 2013)

The fact that our respondents were selling on online marketplace Amazon and keeping up to date with developments in urban areas offers further evidence of the 'diverse networks and flows that criss-cross rural and urban space' (Hoggart, 1990: 43) and illustrates that the distinction between the rural and urban creative economies often is not necessarily a lived reality of rural creative practitioners, other than through a digital divide.

Online Marketing and Promotion

Inadequate internet connectivity only sometimes impacted directly on the creative practices of rural practitioners, but it did influence the way they sought to creatively market themselves. Rural creatives are keenly aware of the opportunities for self-promotion online:

> [With Google analytics] you can analyse which images people are looking at more than others … so it would give me a better feel for the market and how to market myself. (Artist and curator, Aberdeenshire, 2013)

> Well we've got a business Facebook where we will be telling people about events and things that are happening. Facebook works extremely well, as you can imagine, for things like the wedding shows but even LinkedIn … I can actually say that over the last three months I've had three jobs directly linked to LinkedIn. (Marketing business, Aberdeenshire, 2013)

Facebook, Twitter and YouTube are mentioned predominantly as the key sites to upload examples of work and attract clients. Flickr and YouTube act as more formal sites to display a back catalogue of work whilst Facebook and Twitter function in developing networks of practice as well as allowing a visual presence:

> … I've been communicating with people with the likes of Facebook and I've been selling work through Facebook … And on Twitter as well I try to post things that will get people to look at that as well. (Artist and curator, Aberdeenshire, 2013)

Our respondents are also engaging with other social media sites which are tailored to individual creative sectors, such as the music-related site, Reverbnation:

> People have found us directly through Reverbnation … you can get a conversation going back and forth with people … you've kind of got to play the game just to get yourself out there. (Musician, Inverness-shire, 2013)

Not all of the creative practitioners embraced social media – some expressed hesitancy about using social media sites such as Facebook and Twitter:

> I'm still wary of it because some of the things I see that other folk put on it I cringe. (Marketing business, Aberdeenshire, 2013)

Respondents expressed wariness based on factors varying from personal taste to fears about damaging comments being posted; however, most realised that it was an increasingly necessary part of their business practice. Some expressed regret at their own levels of web presence:

> I'm missing out completely on that side of, that way of marketing … you need to have a good visual presence. I do have a Flickr presence … it's a very sort of low key thing but it's there. (Lighting designer, Aberdeenshire, 2013)

Many respondents had their own websites but these could prove problematic either through costs to maintain or through poor connectivity impacting on keeping them up-to-date, with uploading and downloading content, linking to social media, and with time and digital skills limitations. Yet there were obvious benefits from promoting and conducting transactions online:

> [W]ith my art work I believe very much in exhibiting online and being in charge of the marketing of it, especially these days with various galleries on the downturn … it just means I've more control in producing and selling the product, getting feedback from customers. (Artist and bagpiper, Tayside, 2013)

There was some realisation of the value of online promotion, given their rural location:

> If I want to carry on working in Scotland, which I do, I have to make some concessions which is obviously I have to expect to be paid less, for some reason, and to promote myself better … when I was living in London or Nottingham or Sheffield, or whatever, I was part of a network … people who would ring me up

> and say 'are you available for this?' or whatever. There is a network up here but it's so diverse … it's so spread out. (Lighting designer, Aberdeenshire, 2013)

They feel severely disadvantaged by their disability to connect into online networks, or to connect *only* in a patchy way.

Rural creatives are marketing themselves online through social media and business or personal websites, gaining work, sales and contacts globally online. Their visual presence is sometimes low-tech and they express frustration at the limitations to their online transactions, particularly streaming, uploading and real time communication, caused by low connectivity.

The Rural Economy

Our findings suggest that given adequate broadband connectivity, rural creative practitioners are able to contribute to the social and economic sustainability of their communities, particularly through diversification and job creation:

> The nearest comparable businesses would be in Aberdeen so therefore I think it's quite good for the rural economy, the fact that all these people and all my ad-hoc people as well are folk from round about so I think it's bringing quite a lot of employment to an area that there wouldn't have been. (Marketing business, Aberdeenshire, 2013)

> It completely holds back the progress of the business and makes you completely fed up and irritated by the whole thing and puts other people off … It's a huge part of it. Because if you start losing people who are working within the community and stuff, it becomes a dying area – or an area just full of one type of people, it would just be farmers … and it always stops diversity [which] is the best thing for any area. (Children's video producer, Aberdeenshire, 2013)

Yet according to many of the practitioners, if connectivity does not improve, they will face the very real threat of having to move in order to protect their business. Indeed, one of our respondents, the lighting engineer, has already relocated in order to access better broadband connectivity. Others stressed that they had previously been able to function well on the available connectivity, however the connection speeds have failed to improve adequately relative to other parts of the UK:

> If there hadn't been broadband on offer, I don't think we would have considered it. But as the broadband has stood still over seven years … gradually possibilities have diminished. (Music producer with recording studio, Western Isles, 2013)

In main, the respondents stressed their seriousness about making their rural businesses work and the severity of the consequences of lack of digital connectivity;

they did not all represent 'rural lifestyle businesses' in the sense described in the literature. Indeed, some would be better classed as international businesses. None were focusing on a rural or local market exclusively although they expressed their desire to source more work closer to home. Access to improved technology might well permit these broader markets and working relationships – in turn permitting creative industries, perhaps alongside other diverse industries, to flourish in rural, often declining, regions.

Conclusions

Remote rural places differ from urban ones – creative communities in rural locations do not follow Florida's much cited theory of clusters, but instead are characterised by dispersed networks engaged in a portfolio of skills, networked activities and collaborations. Place impacts on rural creatives who need to find ways to communicate and collaborate with remote professional and client networks. This places digital connectivity at the centre of their everyday practices and makes it crucial for their sustainability. From this perspective, urban- and high-tech focused creative policy is not benefitting rural practitioners who primarily seek higher speed broadband internet connections. This situation may change for those who will benefit from BT's recent rollout of undersea cable across many Scottish Islands, however our research has highlighted that digital practices in rural areas are interdependent on other factors (Worth, 2014).

Our findings show that creative practitioners use multiple forms of digital technologies that are crucial to business success. Rural creative practitioners, in turn, are limited by poor connectivity, impacting their businesses in a number of ways, such as their capacity to research their field and remain competitive, learn new skills and therefore innovate, and to collaborate with others further afield. The desire or lack of choice to live in remote Scotland is negotiated through compromises in terms of work and personal life, through everyday tensions between the local and the global and situated practices. Rural creative practitioners use email and social media as an everyday practice, most popularly engaging with Facebook, YouTube, Twitter and Flickr as a means of promotion, making contacts and seeing what competitors are doing. Others access professional networks online as an extension of existing networking activity, such as formal associations for particular professions/sectors. Rural creative practitioners are aware of the need for online presence, but many feel that, due to limited connectivity, their efforts leave a lot to be desired. Further, in order to maintain a professional image, rural practitioners may disguise poor connectivity in a bid to maintain their clients. Creative practitioners often find innovative ways to use the internet to enhance their creative activity. They compensate for issues of poor connectivity through strategies such as: combining on/offline working; working at different times of the day; or connecting elsewhere (e.g. a friend's, a library, or work visits to a city).

Creative practitioners working in remote locations wish to be able to do the same things online, with the same ease, as their competitors.

This chapter has sought to explore in detail the everyday realities or 'relative *permanences*' (Heley and Jones, 2012) of digital practices for rural creative practitioners and the ways in which these are integrated into their creative work and their capacity to contribute to rural economies. It finds that the potential of digital technologies put forward in policy discussions is not being realised for those in remote rural areas, and suggests that CCI policy is aimed at more high-tech sectors and is not able to respond to the low-speed, and sometimes low-tech, realities of rural creative everyday practice. On the whole, our respondents did not reference the national structures offering business support or creative funding, perhaps illustrating that these are not well-known or not viable options for the practitioners we interviewed. However, most rural creative practitioners recognise the benefits and seek to be using more technologies, and compensate for problems caused by inadequate connectivity through a number of creative strategies. Nonetheless, our work supports the notion that without improved broadband infrastructures, rural (particularly remote) regions will struggle to attract – and retain – the diverse industries and economic activities necessary to ensure their resilience in the light of continued global change.

References

Andrew, J. and Spoer, J. (2011) Beyond the creative quick fix: Conceptualising creativity's role in a regional economy, in K. Kourtit, P. Nijkamp and R. Stough (eds) *Drivers of innovation, entrepreneurship and regional dynamics*. New York: Springer, pp. 369–380.

Bain, A. and McLean, H. (2012) The artistic precariat. *Cambridge Journal of Regions, Economy and Society* 6(1), pp. 93–111.

Bell, D. and Jayne, M. (2010) The creative countryside: Policy and practice in the UK rural cultural economy. *Journal of Rural Studies* 26(3), pp. 209–218.

Bowles, K. (2008) Rural cultural research: Notes from a small country town. *Australian Humanities Review* 45, pp. 83–96.

Burns, J. and Kirkpatrick, C. (2008) *Creative industries in the Rural East Midlands: regional study report* http://www.artsderbyshire.org.uk/images/EM_Rural_Creative_Industries_Regional_Report_tcm40–157789.pdf accessed 4 February 2015.

Commission for Rural Communities (2009) *Mind the gap: digital England – a rural perspective* http://www.samknows.com/uploads/CRC.pdf accessed 4 February 2015.

Community Broadband Scotland (2014) *Closing the digital divide* http://www.hie.co.uk/community-support/community-broadband-scotland/news/closing-the-digital-divide.html accessed 4 February 2015.

Crafts Council (2011) *Craft and rural development* http://www.craftscouncil.org.uk/content/files/craft_and_rural_development.pdf accessed 4 February 2015.

DCMS (2001) *Creative industries mapping document* https://www.gov.uk/government/publications/creative-industries-mapping-documents-2001 accessed 4 February 2015.

DCMS (2008) *Creative Britain: new talents for the new economy* http://webarchive.nationalarchives.gov.uk/20080610185304/http://www.culture.gov.uk/images/publications/CEPFeb2008.pdf accessed 4 February 2015.

DCMS (2010) *Britain's superfast broadband future* https://www.gov.uk/government/publications/britains-superfast-broadband-future accessed 4 February 2015.

DCMS (2015) *Creative industries worth £8.8 million an hour to UK economy* https://www.gov.uk/government/news/creative-industries-worth-88-million-an-hour-to-uk-economy accessed 4 February 2015.

DEFRA (2013) *Report for the House of Commons Rural Communities: Government Response to the Committee's Sixth Report of Session* 2013–14.

Donald, B., Gertler, M.S. and Tyler, P. (2013) Creatives after the crash. *Cambridge Journal of Regions, Economy and Society* 6, pp. 3–21.

Duxbury, N. and Campbell, H. (2011) Developing and revitalising rural communities through Arts and Culture. *Small Cities Imprint* 3(1), pp. 111– 122.

European Commission (2005) *European Agricultural Fund for Rural Development* http://europa.eu/legislation_summaries/agriculture/general_framework/l60032_en.htm accessed 4 February 2015.

European Commission (2009) *Report from the Commission to the European Parliament and the Council. Sixth progress report on economic and social cohesion* http://ec.europa.eu/regional_policy/sources/docoffic/official/reports/interim6/com_2009_295_en.pdf accessed 4 February 2015.

European Commission (2010a) *A Digital Agenda for Europe* http://ec.europa.eu/digital-agenda/en accessed 4 February 2015.

European Commission (2010b) *Green Paper: unlocking the potential of cultural and creative industries* Brussels, COM(2010)183 http://ec.europa.eu/culture/our-policy-development/doc/GreenPaper_creative_industries_en.pdf accessed 4 February 2015.

Florida, R.L. (2012) *The Rise of the Creative Class: Revisited.* New York: Basic Books.

Guardian Online (2013) *Superfast broadband for Cameron, but neighbours left in digital slow lane* http://www.theguardian.com/uk-news/2013/dec/01/rural-community-broadband-fund accessed 28 February 2014.

Galloway, L., Sanders, J. and Deakins, D. (2011) Rural small firms' use of the internet: From global to local. *Journal of Rural Studies* 27(3), pp. 254–262.

Gibson, C. (2010) Guest editorial – creative geographies: tales from the "margins". *Australian Geographer* 41(1), pp. 37–41.

Gibson, C., Luckman, S. and Willoughby-Smith, J. (2010) Creativity without Borders? Rethinking remoteness and proximity. *Australian Geographer* 41(1), pp. 25–38.

Hartley, J. (2005) *Creative Industries*. London: Wiley-Blackwell.

Harvey, D.C., Hawkins, H. and Thomas, N.J. (2012) Thinking creative clusters beyond the city: people, places and networks. *Geoforum* 43(3), pp. 529–539.

Heley, J. and Jones, L. (2012) Relational rurals: Some thoughts on relating things and theory in rural studies *Journal of Rural studies*. 28, pp. 208–217.

Helsper, E.J. (2012) A corresponding fields model for the links between social and digital exclusion. *Communication Theory* 22(4), pp. 403–426.

Herslund, L. (2012) The rural creative class: Counterurbanisation and entrepreneurship in the Danish countryside. *Sociologia Ruralis* 52(2), pp. 235–255.

Hoggart, K. (1990) Let's do away with the rural. *Journal of Rural Studies* 6(3), pp. 245–257.

House of Commons Committee of Public Accounts (2013) *The rural broadband programme: Twenty-fourth report of session 2013–14* http://www.publications.parliament.uk/pa/cm201314/cmselect/cmpubacc/474/474.pdf accessed 4 February 2015.

McDonagh, J. (2013) Rural geography I: Changing expectations and contradictions in the rural. *Progress in Human Geography* 37, pp. 712–720.

Northern Ireland Assembly (2013) *Ambitions for the Arts: A five year strategic plan for the arts in Northern Ireland 2013–2018* http://www.artscouncil-ni.org/documents/research/equality/equalityimpactassessment/draft_five_year_strategy.pdf accessed 4 February 2015.

Ofcom (2014) *UK fixed-line broadband performance, May 2014 – The performance of fixed-line broadband delivered to UK residential consumers* http://stakeholders.ofcom.org.uk/market-data-research/other/telecoms-research/broadband-speeds/broadband-speeds-may2014/ accessed 4 February 2015.

Salman, A. (2012) From access to gratification: towards an inclusive digital society. *Asian Social Science* 8(5), pp. 5–15.

Scottish Government (2011a) *Our rural future: The Scottish Government's response to the speak up for Rural Scotland consultation* http://www.scotland.gov.uk/Resource/Doc/344246/0114504.pdf accessed 4 February 2015.

Scottish Government (2011b) *Scotland's Digital Future: A strategy for Scotland* http://www.scotland.gov.uk/Resource/Doc/981/0114237.pdf accessed 4 February 2015.

Scottish Government (2011c) *Growth, talent, ambition – the Government's strategy for the creative industries* http://www.scotland.gov.uk/Publications/2011/03/21093900/0 accessed 4 February 2015.

Scottish Government (2012) *Scottish Government urban rural classification 2011–2012* http://www.scotland.gov.uk/Resource/0039/00399487.pdf accessed 4 February 2015.

Skoglund, W. and Jonsson, G. (2012) The potential of cultural and creative industries in remote areas. *The Nordic Journal of Cultural Policy* 2, pp. 181–191.

Thomas, N.J., Harvey, D.C. and Hawkins, H. (2013) Crafting the region: creative industries and practices of regional space. *Regional Studies* 47(1), pp. 75–88.

Ward, S. and O'Regan, T. (2014) The Northern Rivers media sector: making do in a high-profile rural location. *International Journal of Cultural Policy* online first, pp. 1–18.

Welsh Assembly Government (2010) *Delivering a digital Wales: The Welsh Assembly government's outline framework for action* http://wales.gov.uk/docs/det/publications/101208digitalwalesen.pdf accessed 4 February 2015.

Wojan, T.R., Lambert, D.M. and McGranahan, D.A. (2007) Emoting with their feet: Bohemian attraction to creative milieu. *Journal of Economic Geography* 7(6), pp. 711–736.

Woods, M. (2007) Engaging the global countryside: globalization, hybridity and the reconstitution of rural place. *Progress in Human Geography* 31(4), pp. 485–507.

Worth, D. (2014) *BT completes 250km undersea broadband rollout in Scottish Isles* http://www.v3.co.uk/v3-uk/news/2385004/bt-completes-250km-undersea-broadband-rollout-in-scottish-isles accessed 4 February 2015.

Chapter 8

Libraries and Museums as Breeding Grounds of Social Capital and Creativity: Potential and Challenges in the Post-socialist Context

Monika Murzyn-Kupisz and Jarosław Działek

Introduction

Recent decades have witnessed profound changes in the theory and practice of library and museum management, which has gone beyond the traditional tasks of gathering, storing and displaying exhibits or borrowing books (Vergo, 1989; Black, 2012; Edwards et al., 2013; Svendsen, 2013). In their attempt to develop a more attractive offer, many public cultural institutions strive to increase their visibility and engagement in local communities. Growing numbers of them are redefining their mission in response to the changing expectations of audiences and the public authorities financing them (Building Futures, 2004; Scott, 2003; Sandell and Nightingale, 2012). Contemporary museum visitors expect not only longer opening hours, but also more possibilities for exploration with family and friends, to encounter new people, to engage in educational and leisure activities or even to actively participate in developing museum narratives (Kelly and Gordon, 2002; Kinghorn and Willis, 2008). Similarly, libraries are evolving towards a 'third place' (Elmborg, 2011) where one can not only study or borrow a book, but also be entertained, meet both old friends and new people, and drink coffee (Harris, 2007; Edwards et al., 2013).

At the same time, cultural institutions have been recognized as key elements of cultural industries and the creative economy, as agents that enable the diffusion of ideas and inspirations from artists to other cultural and related industries. In the model elaborated by WIPO (2003) libraries are among the core copyright industries as members of the publishing, literature and photography sectors, while museums belong to the partial copyright industries. According to KEA (2006), activities within cultural industries are divided into the cultural sector, including cultural institutions, and the broader creative sector. The UK's NESTA (2006) locates museums at the intersection of two activity types constitutive for creative industries: i) those delivering and disseminating original artefacts; and, ii) those providing experiences. Libraries are not mentioned directly but they could be included within broadly understood heritage services. Finally, in the concentric circles model of cultural industries (Throsby, 2008) museums and libraries are mentioned among 'other core creative industries' next to and supporting 'core creative arts'.

This recognition notwithstanding, for a long time analysis of the creative sector's contributions to local and regional development has tended to focus on for-profit activities or its immediate, direct economic impact understood as its potential to produce goods and services exported outside a given region, generate profits and employment, attract tourists, or enhance the image of an area for potential investors. In addition, in their assessment of the socio-economic potential of museums and libraries, such studies usually focus on major institutions in large urban centres rather than more peripheral or smaller organisations. Such an approach is not always realistic, nor justified, however. Aside from a few superstar institutions (Greffe, 2011), few museums have the potential to become profit-generating tourist attractions. The ability of libraries to generate direct profits is even more constrained, given that they usually cater free or nearly free of charge to local rather than external audiences. In addition, by focusing on direct, short-term economic impacts, such an approach tends to obscure the other social and economic functions of cultural institutions that are harder to quantify and the more complex roles they may play in local communities and local economies (Scott, 2003).

In the long run museums and libraries might be very important to the development of the creative economy. The presence and activities of such institutions have numerous indirect economic and non-economic effects (Grassier and Grace, 1980). They are institutions where artists and creatives can seek information and inspiration. Some, such as major national libraries, are perceived as 'essential to the ecology of the creative industries' (Brindley, 2008: 33). They not only act as direct customers of creative goods and services but often influence the formation of the tastes of both creatives and customers of creative goods and services. They are also networking venues for different actors in the creative sector, bringing together professionals and amateurs. The spaces in and around cultural institutions may serve as inspiration and backdrops to activities connected with the visual arts, music, TV or film productions. Cultural institutions can also contribute to the unique character of specific quarters, attracting creatives for residence, work and leisure. They fulfil important functions in terms of providing possibilities to satisfy diverse needs (spiritual, aesthetic, leisure, cultural) impacting on quality of life. Is there, however, another dimension of the impact of museums and libraries, perhaps to some extent hitherto overlooked and less obvious, linking the issues of socio-economic development, social engagement, inclusion and the creative inspirations they provide?

In this chapter we explore a less direct way in which libraries and museums influence the creative economy, focusing on their potential simultaneous impact on social capital and creativity. We examine and discuss various aspects of their role as both 'social capital infrastructure' (Warner, 2001) and institutions inspiring creativity, connecting communities to creative endeavours and the creative sector. Rather than looking solely at linkages created between cultural institutions, artists and other creative professionals, then, we are especially interested in strong and weak ties and social trust created between different stakeholders within the local communities where such institutions are located.

The existing scholarship on libraries and museums and their social and creative functions has tended to draw conclusions mainly from case studies based in the Western European or Anglo-Saxon contexts (Bryson et al., 2002; Scott, 2003; Burdett, 2004; Crooke, 2007; Burton and Griffin, 2008; Kinghorn and Willis, 2008; Vårheim et al., 2008; Silverman, 2010; Ferguson, 2012; Johnson, 2012; Sandell and Nightingale, 2012; Edwards et al., 2013; Golding and Modest, 2013; Svendsen, 2013). Our aim is thus to broaden the discussion to include the perspective of post-socialist Central and Eastern Europe. In socialist countries prior to 1989 cultural institutions were generally perceived as a costly Marxist superstructure, useful mainly for the purposes of education and ideological indoctrination, whilst also providing state-sponsored leisure opportunities (Balicki et al., 1973). Aside from ideological undertones, no broader social or economic potential was perceived for them. In the past few years the cultural sector has faced new challenges from post-socialist socio-economic transformation, including opening up to the forces of globalization, privatisation and commercialization, decentralisation resulting in the devolution of most cultural institutions from the central government to local governments, and increasingly neoliberal approaches to the management of the public sector (Varbanova, 2007; Murzyn-Kupisz, 2010; Ilczuk, 2012). This has resulted in a quite intensive discussion on the need to make Polish museums and libraries more economically efficient, with a particular emphasis on finding instrumental rationales for providing them with public funding (Głowacki et al., 2009; Murzyn-Kupisz, 2010; Folga-Januszewska and Gutowski, 2011). Public authorities have begun to assess the potential of many museum institutions, particularly in respect of their capacity to attract large numbers of tourists. Similarly, many libraries are struggling to convince their patrons to maintain support in the age of digitalisation. One of the reasons for this state of affairs is that public officials at both the national and municipal levels, particularly in the first years after Poland's accession to the EU, have tended to rather unreflectively follow the fashion for spectacular cultural flagships dominant in Western Europe since the 1980s, favouring larger scale projects with (at least potential) significant tourist and symbolic potential over modest, smaller-scale investments in more 'basic' cultural infrastructure of local importance. Such flagships also seemed most promising in terms of proving the economic success of EU co-funded investments in their ex-post evaluation. Such a preference may be only partly justified by the fact that the cultural infrastructure inherited from socialism was outdated and to some extent incomplete, particularly in the context of the significant political and social changes (e.g. the lack of museums of modern art and the demand for new museums or museum exhibitions focused on heritage not accepted by the socialist regime). As follows, the first post-socialist cultural investments gave support to major institutions of national or regional importance. Moreover, as no great ideological importance has been attached to the cultural sector since 1989, it has suffered from political marginalization. As observed by Varbanova (2007: 51): 'Throughout the process of democratic transition, culture has always been a low priority for national and regional governments, as other social and economic priorities absorbed most of the political attention and state funding'.

Cultural Institutions and Social Capital

Despite some criticism (Portes, 2000) and not always conclusive empirical evidence (Westlund and Adam, 2010), social capital is nowadays frequently perceived as a crucial development resource (Dale and Newman, 2010; Papaioannou, 2013). Thanks to its ability to mobilize community members to undertake joint action, social capital may be a basis for building other forms of community capital (Edwards et al., 2013). Two main types of social capital are distinguished (Gittel and Vidal, 1998; Putnam, 2000). Bonding social capital refers to the strong ties within family, neighbourhood or friendship groups. The term 'bridging social capital' refers to weaker connections between acquaintances from more varied social backgrounds. Both bonding and bridging social capital are vital to socio-economic development (Woolcock, 1998). The former is thought to be the cornerstone of community integration. The latter reflects the openness of a community and allows access to other economic and cultural resources (Beugelsdijk and van Schaik, 2005). Additionally, a high level of social trust, which, in contrast to personal trust, does not pertain to persons one knows well but also to strangers, is perceived as particularly inspiring to development (Paldam and Svendsen, 2000). Trust reduces concerns about the unpredictable behaviour of others, and gives more space for creativity, innovativeness and entrepreneurial activities (Sztompka, 1998). Moreover, cooperation between individuals translates into capacity for cooperation between public institutions and NGOs which are themselves resources of institutional social capital (Svendsen, 2013). Arguments for greater resources of social capital should also take account of the fact that it is usually developed as the result of long-term social processes, tends to resist attempts at social engineering, and should preferably be enhanced gradually by creating favourable conditions for its development (Wilson, 1997).

In recent years, following discussions around their broader social roles, museums and libraries are increasingly mentioned in the context of social capital (Scott, 2003; Ferguson, 2012; Svendsen, 2013). They offer indoor and outdoor public spaces that are perceived as relatively safe and open (Gurian, 2001; Ferguson, 2012), and as such function as low-intensity meeting places (Aabø et al., 2010). As venues for spending leisure time (Awoniyi, 2001), they help to enhance existing family and friendship links (bonding social capital) (Beaumont and Sterry, 2005; Silverman, 2010). They may also provide opportunities for fostering new relations (bridging social capital), creating networks between persons and groups with similar interests or, conversely, function as spaces for interactions between people who differ significantly from each other, by facilitating informal, unplanned encounters (Johnson, 2012). For example, a library is often:

> … a place where people accidentally run into neighbours and friends, but it is also a place where a substantial proportion report being accidentally engaged in conversations with strangers (… .) a place where users are exposed to 'the other'. (Aabø et al., 2010, 25)

Each museum and library draws and creates around it different communities (Edwards et al., 2013). Museums and libraries may thus function as 'social capital infrastructure' and 'third places' between private, home spaces and study- or workplaces. The social functions of museums and libraries may be further enhanced by the fact that they constitute a relatively dense network of institutions present in both larger and smaller settlements. In addition, they are often housed in major historic landmarks or architecturally attractive buildings located in town or village centres. Library or museum premises serving as public spaces may become important elements of urban regeneration policies alongside renewal of other public fora such as squares and green spaces (Murzyn-Kupisz and Działek, 2013). In smaller settlements, another important element may be the fact that such institutions tend to employ relatively well educated staff, providing communities with potential local leaders (Burton and Griffin, 2008; Svendsen, 2013).

Moreover, museums and libraries have a potential to function as important institutions exerting influence on the knowledge, attitudes and behaviour of a range of communities. They often shape and express local identity by serving as documentation centres and reservoirs of memory for local communities, instrumental in enhancing a sense of local pride and belonging. At the same time, their capacity to collect, display and interpret artefacts linked with non-local cultures may also be important in terms of fostering tolerant attitudes and openness towards diverse social, ethnic or religious groups, which can be enriching culturally and inspiring. Both museums and libraries are thus more and more often perceived as good instruments of social integration, particularly in multicultural societies (Macdonald and Fyfe, 1996; Elbeshausen and Skov, 2004; Audunson et al., 2011; Vårheim, 2011; Edwards et al., 2013). Museums may function as 'agents of social inclusion' on three main levels: through direct cultural activities aimed at marginalized or minority groups; within the framework of broader social undertakings such as cooperation with penitentiary institutions, care homes and programmes addressed to the disabled or the unemployed; and as venues for debate on socially important issues (Sandell, 1998). Consequently they are not merely public spaces in terms of access but also as potential fora for public discourse (Barrett, 2011) and the creation of social trust (Cuno, 2004; Vårheim et al., 2008). Libraries can also reach out to social groups termed as 'information poor' (Britz, 2004), serving as both reservoirs of knowledge and guides pointing to dependable sources of information and helping to overcome digital exclusion (Bertot et al., 2008). As follows, according to some researchers libraries and museums should above all evolve into institutions responding to social challenges (Lang et al., 2006; Simon, 2010). The human capital-enhancing function of cultural institutions through learning, thanks to the introduction of more active and engaging educational forms, may serve not only to improve knowledge or cultural skills but also to hone social and civic competencies (Van Lakerveld and Gussen, 2012). Going even further, Hesmondhalgh (2007: 6) construes individuals, organisations and institutions of cultural industries, including cultural institutions, as 'agents of economic, social and cultural changes'.

Cultural Institutions, Social Capital and Creativity in the Post-socialist Context

Insufficient social capital resources, especially of bridging capital, are often highlighted as one of the main development challenges facing post-socialist countries such as Poland (Boni 2009; Czapiński, 2013; Działek, 2014). In comparison to many West European, especially Nordic countries, the level of bridging capital resources and trust in Poland are rather low, as in other countries of the former Soviet bloc (Mihaylova, 2004; Czapiński, 2013; Papaioannou, 2013). The country's socio-economic development since 1989 has been based mainly on financial and human capital and bonding social capital, and this is reflected in an excessive individualism and 'amoral familiarism' in Polish society. The use of predominantly this type of social capital is derived from the 'social capital of survival' developed during socialist times, i.e. networks of informal contacts that facilitated access to resources and goods not available in the official market (Wedel, 2003; Boni, 2009). Nowadays, as the price of skilled labour increases, further development in Poland will require a greater degree of cooperation, faster exchange of information, and more effective ways of supporting creativity, innovation and entrepreneurship. As a result, one of the main current challenges is building and strengthening trust between citizens and between citizens and the state, and developing networks of links beyond the immediate circle of family and friends (MKiDN, 2013).

The idea that the cultural sector could contribute to the development of social capital is clearly present in most recent major public policy documents in Poland, marking an important change in the perception of cultural institutions and culture since socialist times and the first two decades of the transformation period. The national-level *Social Capital Development Strategy* proposes transformation of cultural institutions into places of encounter and dialogue. They are expected to expand on activities such as 'supporting education other than formal, aimed at cooperation, creativity and social communication', 'supporting cooperation mechanisms between public institutions and citizens', and 'enhancing the role of culture in building social cohesion' (MKiDN, 2013: 7–9). According to the *Long-term National Development Strategy* 'cultural institutions support the construction of 'social capital for development' conducive to cooperation' which makes it necessary to broaden their traditional functions (Boni 2011: 285) (Figure 8.1).

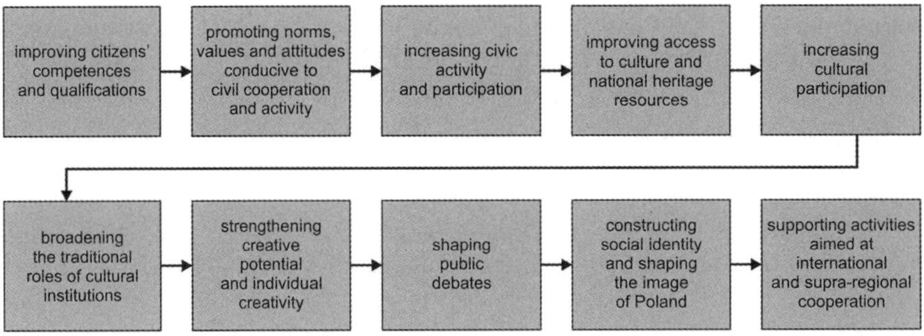

Figure 8.1 Ten aims which make up the main strategic goal of enhancing 'social capital of development' supporting cooperation included in the Long-term National Development Strategy for Poland

Source: compiled from Boni (2011, p. 285)

Analogous tendencies are visible at the regional level. For example, in the *Strategic Programme on Heritage and Leisure Industries* in the Małopolska region, support is foreseen for activities outside museum buildings, projects linked with local communities and involving community art, activities focused on broadening museum audiences to include disadvantaged groups, and campaigns aimed at 'becoming more open to social and creative activities in the museum surroundings' (UMWM, 2012: 83). Similar aims accompany priorities intended for libraries and cultural centres. This is expressed through statements such as:

> [I]t is important that for residents and local communities such institutions function as places of access to knowledge, information and education but also spaces for spending leisure time, talking and participating in culture, [that] they become places of encounters of local artists, cultural and social activists (…) places where artistic, social and cultural projects are conceived and implemented. (UMWM, 2012: 132)

Research Methods

Our main intention is to use the experiences of institutions in post-socialist Poland to consider to what extent and in what ways the potential of libraries and museums to contribute to building and strengthening social capital overlaps with their functions as institutions supporting the development of creativity and the creative economy. The empirical findings presented in this chapter result from a comprehensive study we conducted in museums and libraries in the Małopolska region in southern Poland in 2013. Małopolska is one of 16 Polish regions and has its capital in Krakow. In order to analyse the region's museums and libraries we use a broader theoretical framework pertaining to links between heritage and social capital (Murzyn-Kupisz and Działek, 2013). According to this, several main dimensions of the social-capital related impact of heritage institutions such as museums and libraries may be distinguished (Figure 8.2). The dual purpose of the case study was to test the validity of theoretically defined areas of impact of heritage and heritage institutions on social capital and to arrive at the opportunities and challenges faced by Polish museums and libraries in terms of exerting a positive impact on social capital and creativity.

In the case of museums the analysis took in the internet sites of all active institutions (defined as those open to the general public for at least 5 hours a week) and their branches in the region (N=98), study visits, and in-depth interviews (N=15) with directors and employees. With respect to libraries, which are much more numerous, a questionnaire survey was addressed to all libraries in selected counties in the region (i.e. a sample of municipalities and towns of up to 50,000 residents), which had a response rate of 72 per cent (N=168), and in-depth interviews were conducted in selected institutions (N=4). In addition, an interview was conducted with a regional government official responsible for culture. In the course of the study we made particular note not only of the activities of museum and library institutions with potential for strengthening social capital and inspiring creativity, but also of managers' attitudes towards more active community engagement, cooperation with other actors, inspiration for such activities, and strategies for addressing new challenges faced by the cultural sector.

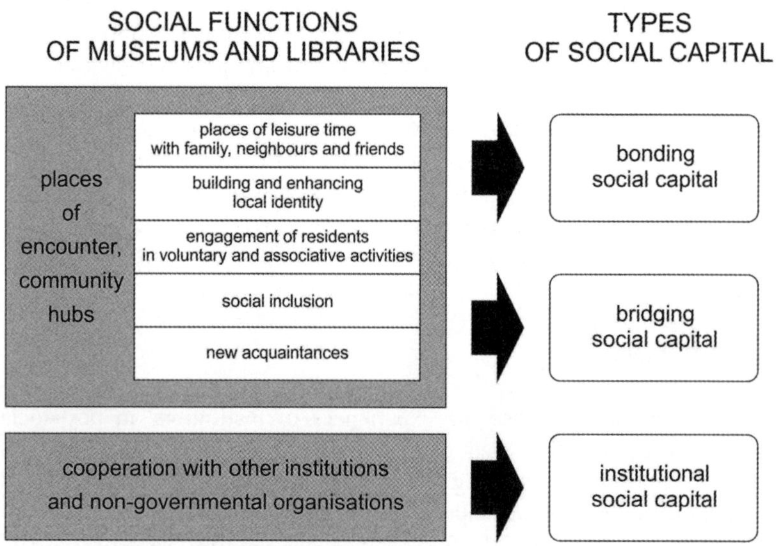

Figure 8.2 Social functions of museums and libraries and their potential impact on social capital

Source: M. Murzyn-Kupisz and J. Działek

Social Capital-building Dimensions and their Links with Creativity and the Creative Sector

The Case of Museums and Libraries in Małopolska

Places of encounter and community hubs

Most libraries in Małopolska, as evidenced by the results of our research and general data on the national level (GUS, 2012), are important meeting places, both for participation in local cultural activities and events, and as leisure venues (Table 8.1). In Poland, as elsewhere, this function of the library is especially important in smaller municipalities, where libraries are often the only place where residents come into direct contact with cultural activities and representatives of creative professions, while also strengthening their social links (Swan et. al., 2013). Most libraries organise exhibitions (often on themes linked to a particular municipality or of art created by local artists) and literary meetings with writers, poets or journalists, attended by residents. Workshops for children, who are the most numerous library users, are frequent, many aimed at encouraging children to engage in creative activities. Other forms of cultural activities, such as film screenings, theatre shows or concerts, are less frequently organised, mainly due to a lack of adequate facilities. Less standard forms of cultural activities in libraries thus tend to take place more often in better equipped library headquarters, usually in towns and larger villages rather than in library branches in smaller settlements. For example in Małopolska, exhibition spaces are available in only half of municipal libraries and just one in five branch libraries. There are dedicated meeting and seminar rooms in every third main library and every eighth library branch. Many events thus have to be organised alongside or at the expense of the library's main functions in corridors or reading rooms, as evidenced in a comment by a library manager in a small town:

> The atmosphere at our meetings is familial, library interiors create a feeling of cosiness, interiors are small and conducive to engaging in conversation (…) there are, however, questions as to whether we should use our spaces for exhibitions or meetings as we have no separate exhibition rooms (…) We really feel the lack of a meeting place. (Library manager, Niepołomice)

Moreover, in view of the poor quality of the library infrastructure inherited from the socialist period, recent investments in libraries have tended to concentrate on core library functions. Arranging spaces to enable a library to undertake broader social and cultural tasks is rarely possible, and usually only happens in the very few institutions housed in thoroughly renovated or completely new buildings.

Figure 8.3 Display of local art at the new main municipal library seat in Myślenice

Source: J. Działek

Table 8.1 Main forms of cultural and educational activities held in public libraries in selected counties of the Małopolska region in 2013 (% of institutions offering such activities)

Cultural and educational activities	Frequency (N=168) in %		
	At least once a month	Every three months or less often	Never
Exhibitions	26.3	46.1	27.5
Workshops for children	24.6	39.5	35.9
Meetings with authors (e.g. evenings with writers or poets)	23.4	58.1	18.6
Celebrations and festivities linked with the seasons, or national and religious holidays	10.3	54.5	35.2
Competitions and contests	9.0	74.7	16.3
Lectures and readings	9.0	45.2	45.8
Thematic events	5.4	28.1	66.5
Film shows	4.2	9.6	86.2
Fairs and flea markets	1.2	10.8	88.0
Theatre events	-	24.6	75.4
Concerts	-	4.8	95.2

Source: Authors' survey, 2013.

Likewise, more and more museums in Małopolska are organising activities in forms other than conventional exhibitions, which are at once creative and conducive to fostering social links. Artistic workshops and competitions, festivals and fairs, and local celebrations of national holidays (Table 8.2) give participants the possibility not only to directly see or learn about artworks or other forms of creative output but also to experience direct involvement in creative endeavours. For example, during the 2013 edition of the 'ETNOmania' festival at the Vistula Ethnographic Park in Wygiełzów, the more than 5,000 participants had a chance to meet over 100 artists and craftspeople, and participate in over 30 workshops, concerts and a fashion show. There are also programmes targeting specific age or family groups, engaging them in creative tasks for which museum collections provide the inspiration. One good example of this is the 'Design Bomb' programme run for children and their parents by the National Museum in Krakow in 2013–2014. During a series of weekend meetings, the attendees learnt about different aspects of design and developed their own design ideas for different objects using museum exhibits as inspiration. In this way museums may also be instrumental in showing how local traditions and historic artefacts (e.g. folk art) can find contemporary uses.

Like their counterparts in libraries, museum managers in Małopolska are increasingly conscious that their institutions need to offer attractive public spaces in order to succeed in drawing visitors. Many institutions offer free access to courtyards and gardens. Some of them stress this feature in promotion activities, while others observe that safe, well maintained green spaces attract local residents, especially on days when there is no entrance fee. In 2013 one in five museums in Małopolska housed a café or a restaurant within its building, while a further third was located near other catering establishments. This was, however, most commonly the case in Krakow. Where such facilities existed in other places in the region it was mainly in larger open-air museums or museums attracting very

Table 8.2 Selected forms of cultural and educational activities taking place in museums in the Małopolska region in 2012

Activity or cultural event type	Number of activities and events
Museum classes	8,346
Film screenings	3,845
Workshops	2,005
Lectures, readings	1,454
Conferences, scholarly meetings	274
Concerts	235
Outdoor events	183
Artistic and historic competitions	120

Source: compiled from GUS Local Data Bank data.

large numbers of visitors. Cafés in museums of a more local character, frequented mainly by residents rather than tourists, in smaller municipalities, are very rare. The issues of designing non-exhibition spaces such as museum shops, cafés and restaurants, educational rooms and workshop spaces usually feature prominently in plans for renovation of existing museum buildings and construction of new facilities. Such investments are, however, most often made by large institutions (national or main regional museums) or in the case of flagship museum projects in major urban centres, particularly Krakow. Smaller institutions, located in peripheral, less populated municipalities are far less likely to benefit from any significant investment. This is not only due to a lack of sufficient local government support or financial problems but in some cases is also linked to unresolved property ownership issues or uncertain tenancy in private buildings.

Problems with regard to making museums places for social gatherings and spending leisure time are also connected to some extent with entrance prices and opening times. Although there was some improvement (in many institutions longer opening hours and more opening days were reported), again this was largely restricted to the regional capital. Most museums continue to remain closed on Mondays and national holidays and very few are accessible after 5 pm, making some groups less likely to visit such institutions (e.g. people in full-time jobs, especially young adults and families with children). Some museums and libraries do experiment with innovative ways of bringing local residents in. For example, in 2013 the Museum in Niepołomice organised a special weekend bus, free of charge, to 'bring the neighbours' to see a temporary exhibition. Problems with weekend and evening access also affect most libraries. Moreover, although still not high by Western European standards, entrance prices to most museums increased significantly after 1989, making their offer relatively expensive from the perspective of some visitors, particularly taking into account the symbolic nature of entrance fees charged in socialist times.

Places of social integration and inclusion

As mentioned earlier, the issue of social inclusion in libraries and museums covers a much broader array of matters than simple physical access. This said, in 2012 only one in four libraries and two in five museums in Małopolska were accessible to people in wheelchairs, while just one in ten libraries and less than a third of museums had interior layouts adjusted to the needs of this group. This is despite the fact that if a major renovation or adaptation of museum or library building is undertaken or a new facility is opened this issue is always taken into account.

Library programmes and activities addressed to various groups in danger of social exclusion are still rare in Małopolska. One in five libraries cooperate with local youth community clubs. A similar proportion of libraries cooperate with local nursing homes. One library in the analysed sample reported cooperation with penitentiary institutions, another mentioned activities aimed directly at an ethnic minority (the Roma). On the other hand, libraries are very important as places teaching digital skills, particularly to children from rural areas and to senior citizens. Such educational activities conducted by libraries may be seen not only as supporting digital inclusion but also equipping some social groups with the skills to both passively (access to content) and actively (creation of content) participate in contemporary culture, access information on cultural events and activities, connect with others, and discuss opinions on creative goods and services. For Polish senior citizens libraries are also places to read paper editions of newspapers and journals, which many of them cannot afford to buy.

Although museum buildings are slightly better appointed to accommodate disabled visitors than libraries, broader, long-term activities serving social inclusion are still also few and far between, conducted by a handful of pioneering institutions. We should note, however, that aside from simple guided tours or museum classes, the most popular form of activities offered explicitly to the disabled are artistic workshops, outdoor painting classes, and sessions for people with impaired hearing and mental disabilities. Some institutions organise exhibitions of art works produced by the disabled as well. Senior citizens are also increasingly targeted as a promising museum audience, though specific programmes addressed to them are mainly organised for already active participants of universities of the third age. Direct involvement of seniors in creative tasks in museums may also include demonstrating or teaching traditional crafts and creative skills to museum visitors or working as volunteers reading books to children (important from the perspective of fostering intergenerational links). Activities linking other 'socially sensitive' groups, such as prisoners, to the creative economy are rarer still, although some good practices are emerging. Two museums in the region not only conduct lectures and courses for convicts both in museums and directly in prisons, but also involve them in preparations for exhibitions, conservation work or archaeological excavations. A unique initiative in Chrzanów is the 'Review of Prison Art' – an annual exhibition of artworks created by convicts. Due to the largely mono-ethnic structure of present-day Polish society programmes addressed to ethnic minorities are not common, though programmes referencing the region's historic multiculturalism are increasingly visible in museum activities (Murzyn-Kupisz, 2015). These include participation by museums in preparation of festivals, memorial activities and artistic competitions focused on a given ethnic minority (the Jews, the Roma, the Lemkos). For example the museum in Tarnów is actively involved in cooperation with the Roma community in the organisation of the annual International Roma Memorial Caravan and the International Roma Poetry Contest. Such involvement may likewise consist in providing space for cultural events organised by a given minority (e.g. the Lemkos in Nowy Sącz). Involving local communities or different social groups in curatorial work is still extremely rare in Poland, as confirmed by our research.

Sources of local identity and pride

Museums and libraries act as key institutions in terms of maintaining and enhancing local identity on the basis of local artistic traditions and contemporary artistic output. Practically all the museums in Małopolska collect artefacts linked with regional culture, history and traditions, focus on them in permanent and temporary exhibitions, conduct research on local past, and publish, collect and promote books on local themes. The subject of their endeavours are often various genres of local creative output such as objects of fine and applied art, traditional crafts, literary achievements or festive traditions. Moreover, many museums currently offer workshops on areas such as regional handicrafts, folk art and traditional artistic genres (e.g. painting on glass, lacemaking, paper flower making, embroidery, wood carving), organised as one-off events or longer courses. In some cases, developments to museum premises have focused on this aspect of their activities in particular (cf. the new educational building of the regional museum in Rabka Zdrój). Museums are also often the main organisers or co-organisers of annual celebrations and outdoor cultural events focused on local traditions during which professionals and amateurs present their work (e.g. the Historical Museum of the City of Krakow is the main organiser of the traditional Krakow Christmas nativity scene competition). In addition, many museums conduct research into some expressions of local creativity, for example creating databases of artists, designers and craftspeople (e.g. in Kęty). They also encourage creativity based on local themes through competitions and contests (e.g. literary, drawing or photography). In addition museums may educate their audiences on the potential of their collections for the creative industries. An example is the 'Etnodizajn' festival, organised in 2009–2010 by the Ethnographic Museum in Krakow with the aim of informing the general public and artists about the ways ethnographic collections may provide inspirations for contemporary design (Klekot, 2010) including furniture, fashion, toys and street furniture. It also resulted in the creation of an ethno-inspired design pattern book and a database of artists and designers using such motifs.

Similarly, most libraries (over 90 per cent in the sample) collect publications on local themes. They frequently conduct research into local history and traditions, of which documenting local artistic traditions often forms an important part (i.e. information on the life and work of local writers, painters, literature based on local motifs). They may also publish the results of this research and convey them to the public. Exhibitions presenting different aspects of local history, heritage, arts and crafts are organised by fewer than one-third of the libraries in the Małopolska region, however.

Figure 8.4 Children using a playground installed in Krakow as part of the 'Etnodizajn' festival organised by the Ethnographic Museum in Krakow

Source: A. Majewski courtesy of the Ethnographic Museum in Krakow

Initiators and sites of cooperation with NGOs and volunteers

Aside from the fact that a few museums in the region are run directly by third-sector institutions and may focus on the achievements of a specific prominent artistic figure (e.g. the museum of the poet Jan Kasprowicz in Zakopane), many museum and library institutions also offer scientific and organisational support to NGOs. If possible libraries and museums provide (usually free of charge) meeting, exhibition and performance space for local creative associations such as painters' associations, choirs, poetry groups, literary discussion clubs, photography clubs, theatre ensembles and historic re-enactment groups. They may also provide space for the activities of informal interest groups ranging from knitting and lace-making to board games enthusiasts, and rent or provide free of charge space for artistic workshops conducted by NGOs (e.g. theatre associations) and exhibition space for shows organised by associations focusing on specific issues (e.g. Jewish heritage). Such cooperation also extends to exchange of experiences and help in organisation of cultural events with partners such as voluntary fire brigades, associations of village women, local action groups, heritage societies, local collectors, religious or other cultural institutions, and public authorities. Over one-third of the libraries in

Małopolska declared cooperation with local non-governmental institutions. Such cooperation not only leads to organisation of artistically and culturally significant endeavours but also contributes to building institutional social capital and trust between cooperating partners.

Only one in six libraries in Małopolska use voluntary help. Volunteers do contribute significant involvement, however, in non-core library services and social functions such as workshops for children and seniors, reading to children, delivering books to the disabled, or help in digitalization of library collections. Likewise, volunteering is still a relatively new and infrequent phenomenon in museums although where it is used, it is often in connection with their non-standard functions (e.g. supplementary research, organisation of creative workshops and cultural events).

Institutions cooperating with artists, creative entrepreneurs and organisations
Libraries and museums are important places and institutions connecting artists and other creatives with their audiences and potential customers as well as with other creatives and cultural institutions. Above all they help artists and other creatives to become more visible and familiar to their audiences by providing them with venues for various cultural events. Participation in cultural undertakings may also bring together and create links between the artists and other creatives involved, as well as build trust between artists, audiences and cultural institutions. Lastly, museums can inspire and make it possible for artists to participate in important social debates through their art.

Conclusions

A new outlook on museums and libraries and a new understanding of their social roles goes beyond their basic statutory functions or their traditional roles as preservers, researchers and disseminators of knowledge on collected artifacts or books. The survey of such institutions in the Małopolska region revealed a significant degree of synergy between their social capital-building and creativity-enhancing functions. The role of museums and libraries in connecting communities with the creative economy might be understood as inspiring professional and amateur artistic creativity, providing opportunities for encounters with artists and other creative professionals, and increasing awareness of and stimulating demand for creative products and services. This to a large extent overlaps with the social capital-building functions of such institutions (Table 8.3). Support for their development by public authorities may thus contribute to achieving two complementary policy aims.

Table 8.3 Compatibility of the social capital-building and creativity-enhancing functions of museums and libraries

Social capital-building dimensions	Links between social capital and creativity
Places of encounter and community hubs	**Places for meeting during cultural and creative events and activities** • Places of encounter between persons working in creative and non-creative occupations, offering possibilities to meet and interact with creatives and their work • Possibility to strengthen family and friendship links and establish new connections with strangers while participating in creative activities
Places of social integration and inclusion	**Social inclusion through artistic and creative activities** • Places where diverse social groups experience a broad variety of historic and contemporary creative output • Institutions engaging diverse social groups in artistic and creative activities by providing space, instruction and support for tasks and activities requiring creativity • Venues where they can present their creative efforts (literary, visual, etc.) • Places for exchange of creative skills between different social and age groups • Places teaching and developing digital skills necessary for participation in contemporary culture • Institutions inspiring artists to participate in social debate through their art • Institutions engaging diverse social groups in curatorial activities and creation of museum narrations
Sources of local identity and pride	**Strengthening and maintaining local identity through local creative output** • Places documenting, collecting and presenting local cultural and creative output, professional and amateur, past and present • Places of transmitting and sharing local creative skills and traditions
Initiators and sites of cooperation with NGOs and volunteers	**Institutions supporting and cooperating with creative associations, groups and volunteers** • Institutions providing meeting, exhibition and performance space for local creative associations and groups • Institutions cooperating with creative associations and groups and providing them with help in organisation of cultural events and endeavours • Institutions enabling volunteers to acquire and transmit creative skills and knowledge on the organization of cultural endeavours • Building institutional social capital between the public sector and civil society • Strengthening social trust between cooperating partners
Institutions cooperating with artists, creative entrepreneurs and organisations	**Places/institutions linking artists and creatives with their audiences and clients, other creatives and cultural institutions** • Institutions fostering links between artists and creatives • Institutions enabling artists to communicate with their audiences • Building trust between artists and institutions • Building institutional social capital and trust between cultural institutions of different types

Source: Authors' analysis

As they develop their social and creative functions, however, museums and libraries face many new challenges. Some of these reflect trends and problems experienced by such institutions all over the world. Others are to a large extent linked to the specific historic, cultural, political and economic context in which particular institutions function, in this case the specific post-socialist context.

First of all, museums and libraries are constantly forced to balance between new and traditional functions vital to their specific institutional identity. In order to fulfil growing expectations with respect to their impact, it is usually necessary to introduce many changes in their operations, by making infrastructural improvements, reshaping public spaces around them, implementing new models of management, diversifying employees' skills, allowing for new forms of activities, and changing their image. For example, disabled access to a museum or library's offer no longer stops at physical access to buildings, but extends to making their content more accessible (e.g. braille, audio guides) as well as to including material relevant to the disabled community in collections (Sandell et al., 2010). A lot depends on the ability and willingness of managers and employees to develop a diverse offer of activities and services going beyond their core functions. Availability of indoor and outdoor spaces where such activities could take place is likewise a key issue.

In recent years many museum and library institutions in Małopolska have made positive changes in terms of creating new spaces and undertaking new forms of activities which may be conducive to strengthening existing and developing new social ties. The emergence of a new generation of museum and library managers familiar with international trends and new investments, often co-financed by the EU, is frequently a turning point in rethinking the organisational and functional models of institutions. On the other hand, as a reflection of a global tendency, in the case of museums huge differences are observed between the dynamic changes visible in larger urban centres (especially Krakow, where in addition to the large regular and student population of the city, tourists make up a significant proportion of museum audiences) and major museum institutions, and the stagnation or even decline experienced by many smaller museums or museum branches in less populated, less wealthy, peripheral municipalities. While many of the former are seeing significant transformations and benefiting from new investments and activities, this trend is much rarer among the latter and is also to a large extent linked with the attitudes of local authorities and the financial situation of the municipalities, NGOs or private owners managing such smaller museums. The situation is equally complex in the case of the much more numerous libraries, although the economic capacity of local governments coupled with opinions of local political leaders on the usefulness and potential of such institutions have a decisive impact on their willingness to support libraries and invest in them. As in the case of museums, a similar dichotomy exists between major institutions or main libraries and smaller, peripheral branches, which often suffer from underinvestment despite the fact that they 'serve a strategic role in extending public services to residents that may be hard to reach by other means' (Swan et al., 2013, p. 9).

Another threat, as mentioned above, is that even if local officials do start to espouse the rhetoric of culture, the creative industries and development, they are most likely to be keen on providing support to major, spectacular cultural investments or facilities perceived as desirable by tourists and selected social groups. This may lead to a growing gap between large cities and peripheral municipalities as well as marginalization and insufficient development of locally oriented cultural infrastructure with little tourist or commercial potential despite its great social importance. There is a further risk that, following completion of many major, EU co-funded investments, a growing share of the public budget for culture will go on supporting the costs of their maintenance and day-to-day operations, making it difficult for smaller players in the cultural sector to obtain additional funding for infrastructural improvements and development.

The problems with initiating more social capital-building and creative activities in museums and libraries in Poland, however, are not only related to infrastructural or financial constraints but are also linked with institutional traditions (e.g. a focus on school groups, tours and lectures in museums), negative stereotypes of such institutions which persist among some public officials and the general public, and the rather limited demand of visitors for such broader services. The first issue is probably easier to solve. For instance, over the past decade many museums have developed their offer to reach non-traditional participants such as children visiting museums on an individual basis, family groups, third-age groups, and the disabled. The second matter requires broader attitudinal changes. Museum managers in Małopolska report that most visitors prefer passive visiting rather than active involvement in creative activities in a museum setting. For example, most parents are happy to leave their children in the care of museum staff rather than actively engage in museum workshops. Similarly, attitudes among local (regional) government officials change very slowly. Few of them are aware of and convinced of the multidimensional, long-term potential of cultural institutions in terms of their contribution to local development, though some notable exceptions do exist. Low educational and cultural aspirations of the general public and consumption-oriented life-styles coupled with insufficient cultural education also result in poor cultural competencies to become frequent clients (and beneficiaries) of the broader offer of cultural institutions.

The ability of museums and libraries in Małopolska to sustain and develop social capital and stimulate creativity is, moreover, to a large extent shaped by the post-socialist context in which these institutions function. The discussion on the broader functions of cultural institutions in Poland began later than in Western Europe. The public cultural sector received minimal funds for investments in the first years of transformation, and consequently focused on maintenance of existing institutions and their buildings rather than their development. Most newer construction and renovation projects were undertaken only in the past decade which also begs the question of whether some changes in museums and libraries stem from grassroots initiatives responding to genuine community needs or rather from the simple desire of some cultural managers and public authorities to be 'up to date', follow global or European trends, or attract external funding. Following the devolution of responsibilities to self-governing regions, counties and municipalities (Murzyn-Kupisz, 2010) many of them inherited dilapidated, outdated cultural infrastructure that was difficult and extremely costly to adapt to present-day standards. Another distinctly post-socialist feature of the functioning of museums and libraries in Poland is the fact that the situation of many institutions is rather unstable, making it very hard for them to plan anything in the long term or conduct any investments due to unresolved ownership issues or tenancy in private premises reclaimed by private owners after 1989.

Despite long traditions, the third sector in Poland is still rather weak; many of its institutions closed down or their activities were limited in the socialist period, while the new NGOs which have emerged in the last 25 years still need to attract more members and funds and obtain experiences. On the other hand, some traditional NGOs which enjoyed significant state patronage in socialist times and were founders and managers of some cultural institutions are now experiencing significant financial problems due to the withdrawal of most state support and rapidly decreasing membership numbers. The Polish Tourist and Sightseeing Society (PTTK), which still runs eight museums in Małopolska, is one such example. The slow development of contemporary volunteering and underdeveloped private patronage for culture are not helpful either. The new economic elites are very rarely willing to support museum and library initiatives, while, particularly among older generations, negative attitudes towards voluntary work may persist due to the fact that prior to 1989 most Polish citizens were obliged to perform unpaid work 'in the service of the community' on a regular basis. In addition to financial constraints, many public cultural institutions lack sufficient marketing abilities to be able to compete for customers' leisure time and leisure-related spending. Numerous, skilfully advertised commercial establishments such as theme parks or shopping and entertainment centres, are new and increasingly popular in the Polish context as part of broader, significant lifestyle changes and consumption tendencies emergent in the country in the past few decades.

On the other hand, increasing awareness of good practices and personal experiences of library and museum visitors of similar services abroad, as well as the intensifying critique of consumption-led lifestyles and passive cultural consumption, may increase the expectations and willingness of some visitors and users to be more involved in museum and library activities, bringing benefits in terms of both social capital and creativity to individuals and communities alike.

Acknowledgments

This chapter is based on work carried out as part of a research project on 'Museum institutions: a cultural economics perspective' conducted in 2012–2016. The project is financed with a grant awarded by the National Science Centre, Poland (award number DEC-2011/03/B/HS4/01148). The authors would like to sincerely thank all the museum and library managers and employees who took the time to complete the questionnaires and give interviews for this project.

References

Aabø, S., Audunson, R. and Vårheim, A. (2010) How do public libraries function as meeting places? *Library & Information Science Research* 32, pp. 16–26.

Audunson, R., Essmat, S. and Aabø, S. (2011) Public libraries: a meeting place for immigrant women? *Library & Information Science Research* 33, pp. 220–227.

Awoniyi, S. (2001) The contemporary museum and leisure: recreation as a museum function. *Museum Management and Curatorship* 19(3), pp. 297–308.

Balicki, S.W, Kossak, J. and Żuławski M. (1973) *Cultural Policy in Poland.* Paris: UNESCO.

Barrett, J. (2011) *Museums and the Public Sphere* Chichester: Wiley-Blackwell.

Beaumont, E. and Sterry, P. (2005) A study of grandparents and grandchildren as visitors to museums and art galleries in the UK. *Museum and Society* 3(3), pp. 167–180.

Bertot, J.C., McClure, C.R. and Jaeger, P.T. (2008) The impacts of free public Internet access on public library patrons and communities. *Library Quarterly* 78(3), pp. 285–301.

Beugelsdijk, S. and van Schaik, T. (2005) Social capital and growth in European regions: an empirical test. *European Journal of Political Economy* 21(2), pp. 301–324.

Black, G. (2012) *Transforming Museums in the Twenty-first Century.* London: Routledge.

Boni, M. ed. (2009) *Polska 2030. Wyzwania rozwojowe.* Warszawa: Zespół Doradców Strategicznych Prezesa Rady Ministrów.

Boni, M. (2011) *Polska 2030. Trzecia fala nowoczesności. Długookresowa Strategia Rozwoju Kraju. Projekt. Część II.* Warszawa: Kancelaria Prezesa Rady Ministrów.

Brindley, D. (2008) Culture, creativity and research. *New Review of Academic Librarianship* 14, pp.17–35.

Britz, J. (2004) To know or not to know: A moral reflection on information poverty. *Journal of Information Science* 30, pp. 192–204.

Bryson, J., Usherwood, B. and Streatfield, D. (2002) *Social Impact Audit for the Southwest Museums Libraries and Archives Council.* Sheffield: The University of Sheffield.

Building Futures (2004) *21st Century Libraries. Changing Forms, Changing Futures.* London: Commission for Architecture and the Built Environment.

Burdett, R. (2004) *Museums and Galleries: Creative Engagement.* London: National Museum Directors' Conference.

Burton, C. and Griffin, J. (2008) More than a museum? Understanding how small museums contribute to social capital in regional communities. *Asia Pacific Journal of Arts and Cultural Management* 5(1), pp. 314–332.

Crooke, E. (2007) *Museums and Community: Ideas, Issues and Challenges*. London; New York: Routledge.

Cuno, J. ed. (2004) *Whose Muse? Art Museums and the Public Trust*. Princeton: Princeton University Press.

Czapiński, J. (2013) Social capital. Social Diagnosis 2013. The Objective and Subjective Quality of Life in Poland [Special issue]. *Contemporary Economics* 7, pp. 296–315.

Dale, A. and Newman, L. (2010) Social capital: a necessary and sufficient condition for sustainable community development? *Community Development Journal,* 45(1), pp. 5–21.

Działek, J. (2014) Is social capital useful for explaining economic development in Polish regions? *Geografiska Annaler: Series B, Human Geography* 96(2), pp. 177–193.

Edwards, J.B., Robinson M.S. and Unger K.R. (2013) *Transforming Libraries, Building Communities: The Community-Centered Library*. Plymouth: Scarecrow Press.

Elbeshausen, H. and Skov, P. (2004) Public library in a multicultural space: a case study of integration in local communities. *New Library World* 105(3/4), pp. 131–141.

Elmborg, J. (2011) Libraries as the spaces between us. Recognizing and valuing the third space. *Reference & User Services Quarterly* 4, pp. 338–350.

Ferguson, S. (2012) Are public libraries developers of social capital? A review of their contribution and attempts to demonstrate it. *The Australian Library Journal* 61(1), pp. 22–33.

Folga-Januszewska, D. and Gutowski, B. ed. (2011) *Ekonomia muzeum*. Kraków: Universitas.

Gittell, R. and Vidal, A. (1998) *Community Organizing: Building Social Capital as a Development Strategy*. Thousand Oaks, CA: Sage.

Głowacki, J., Hausner, J., Jakóbik, K., Markiel, K., Mituś, A. and Żabiński M. (2009) *Finansowanie kultury i zarządzanie instytucjami kultury*. Kraków: Uniwersytet Ekonomiczny w Krakowie, MSAP.

Golding, V. and Modest, W. ed. (2013) *Museums and Communities. Curators, Collections and Collaboration*. London: Bloomsbury.

Grassier, R.S. and Grace, R. (1980) The economic functions of nonprofit enterprise: the case of art museums. *Journal of Cultural Economics* 1, pp. 19–32.

Greffe, X. (2011) The economic impact of the Louvre. *The Journal of Arts Management, Law and Society* 41, pp. 121–137.

Gurian, E.H. (2001) Function follows form: how mixed-used spaces in museums build community. *Curator* 1, pp. 97–112.

GUS (2012) *Kultura w 2011 r. Culture in 2011*. Warszawa: Główny Urząd Statystyczny (GUS).

Harris, C. (2007) Libraries with lattes: the new third place. *Aplis* 20(4), pp. 145–152.

Hesmondhalgh, D. (2007) *The Cultural Industries*. London: Sage.

Ilczuk, D. (2012) *Poland: Cultural Policy Profile* http://www.culturalpolicies.net/web/poland.php accessed 10 November 2014.

Johnson, C.A., 2012. How do public libraries create social capital? An analysis of interactions between library staff and patrons. *Library & Information Science Research* 34, pp. 52–62.

KEA (2006) *The Economy of Culture in Europe. Study prepared for the European Commission* http://www.keanet.eu/ecoculture/studynew.pdf accessed 10 September 2014.

Kelly, L. and Gordon, P. (2002) Developing a Community of Practice: Museums and Reconciliation in Australia, in Sandell, R. (ed.) *Museums, Society, Inequality*. London: Routledge, pp. 153–174.

Kinghorn, N. and Willis, K. (2008) Measuring museum visitor preferences towards opportunities for developing social capital: an application of a choice experiment to the discovery museum. *International Journal of Heritage Studies* 14, pp. 555–572.

Klekot, E. (2010) The seventh life of Polish folk art and craft. *Etnoloska Tribina* 30, pp. 71–85.

Lang, C., Reeve, J. and Woollard, V. ed. (2006) *The Responsive Museum. Working with Audiences in the 21st Century*. Aldershot: Ashgate.

Macdonald, S. and Fyfe, G. ed. (1996) *Theorizing Museums: Representing Identity and Diversity in a Changing World*. Cambridge, Mass.: Blackwell.

Mihaylova, D. (2004) *Social capital in Central and Eastern Europe. A Critical Assessment and Literature Review*. Budapest: Central European University.

MKiDN (2013) *Strategia Rozwoju Kapitału Społecznego 2020*. Warszawa: Ministerstwo Kultury i Dziedzictwa Narodowego (MKiDN).

Murzyn-Kupisz, M. (2010) Cultural policy at the regional level: a decade of experiences of new regions in Poland. *Cultural Trends* 19(1–2), pp. 65–80.

Murzyn-Kupisz, M. (2015) Multicultural heritage of Galicia in the contemporary museum activities in southern Poland, in K. Jagodzińska, J. Purchla (ed.) *The Limits of Heritage. 2nd Heritage Forum of Central Europe*. Krakow: ICC, forthcoming.

Murzyn-Kupisz, M. and Działek, J. (2013) Cultural heritage in building and enhancing social capital. *Journal of Cultural Heritage Management and Sustainable Development* 3(1), pp. 35–54.

NESTA (2006) *Creating growth. How the UK can create world class creative business* www.nesta.org.uk/library/documents/Creating-Growth.pdf accessed 10 September 2014.

Paldam, M. and Svendsen, G.T. (2000) An essay on social capital: looking for the fire behind the smoke. *European Journal of Political Economy* 16(2), pp. 339–366.

Papaioannou, E. (2013) *Trust(ing) in Europe? How Increased Social Capital can Contribute to Economic Development*. Brussels: Centre for European Studies.

Portes, A. (2000) The two meanings of social capital. *Sociological Forum* 1, pp. 1–12.

Putnam, R.D. (2000) *Bowling Alone: The Collapse and Revival of American Community*. New York: Simon & Schuster.

Sandell, R. (1998) Museums as agents of social inclusion. *Museum Management and Curatorship* 4, pp. 401–418.

Sandell, R., Dodd, J. and Garland-Thomson, R. ed. (2010) *Re-presenting Disability: Activism and Agency in the Museum*. London: Routledge.

Sandell, R. and Nightingale, E. (2012) *Museums, Equality and Social Justice*. Abingdon, Oxon; New York: Routledge.

Scott, C. (2003) Museums and impact. *Curator: The Museum Journal* 46(3), pp. 293–310.

Silverman, L.H. (2010) *The Social Work of Museums*. London: Routledge.

Simon, N. (2010) *The Participatory Museum*. Santa Cruz: Museum 2.0.

Svendsen, G.L.H. (2013) Public libraries as breeding grounds for bonding, bridging and institutional social capital: the case of branch libraries in rural Denmark. *Sociologia Ruralis* 53(1), pp. 52–73.

Swan, D.W., Grimes, J. and Owens, T. (2013) *The State of Small and Rural Libraries in the United States*. Research Brief No. 5, Washington: Institute of Museum and Library Services

Sztompka, P. (1998) Trust, distrust and two paradoxes of democracy. *European Journal of Social Theory* 1(1), pp. 19–32.

Throsby, D. (2008) The concentric circles model of the cultural industries. *Cultural Trends* 3, pp. 147–164.

UMWM (2012) *Program Strategiczny Dziedzictwo i Przemysły Czasu Wolnego*. Kraków: Urząd Marszałkowski Województwa Małopolskiego (UMWM).

Van Lakerveld, J. and Gussen, I. ed. (2012) *Nabywanie kluczowych kompetencji poprzez edukację na rzecz dziedzictwa kulturowego* http://the-aqueduct.eu/download/Aqueduct-Manual_PO.pdf accessed 10 September 2014.

Varbanova, L. (2007) The European Union enlargement process: culture in between national policies and European priorities. *Journal of Arts Management, Law and Society* 1, pp. 48–64.

Vårheim, A. (2011) Gracious space: library programming strategies towards immigrants as tools in the creation of social capital. *Library & Information Science Research* 33, pp. 12–18.

Vårheim, A., Steinmo, S. and Ide, E. (2008) Do libraries matter? Public libraries and the creation of social capital. *Journal of Documentation* 64(6), pp. 877–892.

Vergo, P. ed. (1989) *The New Museology*. London: Reaktion Books.

Warner, M. (2001) Building social capital: the role of local government. *Journal of Socio-Economics* 30, pp. 187–192.

Wedel, J.R. (2003) Dirty togetherness: institutional nomads, networks, and the state-private interface in Central and Eastern Europe and the former Soviet Union. *Polish Sociological Review* 2, pp. 139–159.

Westlund, H. and Adam, F. (2010) Social capital and economic performance: a meta-analysis of 65 studies. *European Planning Studies* 18, pp. 893–919.

Wilson, P.A. (1997) Building social capital: a learning agenda for the twenty-first century. *Urban Studies* 5–6, pp. 745–760.

WIPO (2003) *Guide on Surveying the Economic Copyright-Based Industries*. Geneva: WIPO

Woolcock, M. (1998) Social capital and economic development: toward a theoretical synthesis and policy framework. *Theory and Society* 2, pp. 151–208.

Chapter 9

Cross Intermediation? Policy, Creative Industries and Cultures Across the EU

Paul Long and Steve Harding

Introduction

This chapter contributes to the holistic approach to intermediation taken in this collection by exploring how people and places are connected into the creative economy at a European level. Its concerns are framed around *Cross Innovation*, a project supported by EU funds and which promotes policies, generates research and initiates practical measures seeking productive interactions, or 'spill-overs', between the creative industries and other sectors (See: cross-innovation.eu). The project is founded on a partnership consisting of 11 metropolitan centres across Europe: Birmingham, Amsterdam, Rome, Berlin, Tallinn, Warsaw, Vilnius, Stockholm, Linz, Lisbon, and Pilsen. As we outline below, *Cross Innovation* is therefore representative of a range of initiatives that prompt questions regarding the role of culture and creative industries as part of the 'European project'.

The conjunction of policy and practice that identifies and nurtures what is termed 'best practice' in *Cross Innovation* offers an empirical means of understanding the role of cultural intermediaries. The process of intermediation is one in which the nature of the cultural as an aspect of the 'economic imaginary' is understood, made meaningful and manifest in local policy directives, concrete activities and objects (O'Connor, 2015; Taylor, 2015). Our approach is one that thus responds to Calvin Taylor's suggestion that there is a 'need to look at the processual character of intermediation per se' (Taylor, 2015: 370). The authors themselves have a role in the *Cross Innovation* project. This instance is then a valuable opportunity for reflexivity with regards our own role in intermediation processes in the context of this collection's problematization and critical evaluation of the mobilising concepts of projects such as *Cross Innovation*.

We recognize how this is a project that evinces a set of normative ideas and conventions about the cultural economy and the initiatives, structures and conventions of the sector and their application through a variety of mechanisms. The questions underpinning this reflection ask: what are the dynamics of policy in practice in this project that assembles different national cultures, institutions, organizational practices and conventions? What ideas emerge about the creative sector and its relationship with a wider economy and cultural imaginary? On what terms are specific concepts and practices translated into practice? Some of the answers we provide evince familiar tensions between the apparent economic instrumentalism of creative industries and a wider sense of the value of culture as integral to identity. This is illustrated across the diverse cultures of Europe, and the range of creative sectors represented here, of the ambiguities and challenge of enlisting creativity as *industry*, and its apparent promise as a site of potential redemption for the EU economy. At the centre of all of this is the intermediary – understood in terms of individuals, institutions and processes of intermediation – negotiating and making sense of the meaning of policy and its implementation amidst the diversity of the EU.

European Culture and the Creative Economy

As Terry Flew (2012) observes, Europe is a rich site for the consideration of creative industries policy. Its development involves a move from a rarefied sense of culture to a prioritizing of economic objectives. While there is a varied continental lineage of cultural policy to acknowledge, there has certainly been an efflorescence of pronouncements and activity in this field since the signing of the Treaty of Maastricht of February 1992. As Psychogiopoulou (2008) notes, the formulation of Article 151 of the treaty marked the moment in which the EU was afforded a remit and direct powers and a remit for intervention in cultural work. This statement can be seen to provide the fount of the ideas that have been examined and followed through in a variety of instruments such as the Green Paper of 2010. In turn, ideas at EU level have been informed by developments and approaches elsewhere such as the establishment in the UK of the DCMS in 1997, the work of Nesta on R&D for the arts (Bakhshi and Throsby, 2010; 2012) or the UNESCO convention on the protection and promotion of the diversity of cultural expressions of 2005. In light of the distinct character of its various member states, that the EU involves itself in this field at all invites some reflection on the way in which policy has something to say about culture as identity and as an autonomous and disinterested field defined by aesthetic considerations as well as the intertwined relationship of both with interventions in the governance of the creative economy (Barnett, 2001; Banus, 2002; Sassatelli, 2002; Tzaliki, 2007; Gordon, 2010).

In terms of European identity, Article 151 as originally drafted (and modified by the Treaty of Amsterdam), was founded on a desire to preserve and promote diversity. It was a means of safeguarding against the homogenization of the richness of EU cultural heritage and character. For Psychogiopoulou, while there is a pragmatic and flexible looseness about terminology that services the diversity of member states, there is also considerable ambiguity in what she labels Article 151's 'cautious language' (Psychogiopoulou, 2008: 27). In this argument, this language gives little firm guidance for how a concept of culture might be formulated for a European political and economic entity. Article 151 expresses a spirit of benign cooperation yet there are potentially anxiety-inducing expressions about European identity and duties such as the aim of 'respecting […] national and regional diversity and at the same time bringing the common cultural heritage to the fore' (European Community, 1997), As Psychogiopoulou (2008: 28) concludes, 'To talk in terms of a single European historical culture may quickly lead to unsustainable generalization, precisely because several historical narratives of Europe are possible'.

It is important to note here that there is no sense in which EU policy is *prescriptive* about what culture should be. Rather, Article 151 sought to facilitate a shared cultural environment, aiding social cohesion and understanding in a period of tension and dispute, an aim conducive to nurturing a sense of a distinctly European community in tandem with more specific economic and political objectives. Of course, culture may exclude, as much as it includes, yet this view of the aim of policy is one that seeks to engage its citizens as cultural consumers *and* agents, extending to the promotion of the 'socio-economic conditions required for an active engagement in cultural production' (Psychogiopoulou, 2008: 34).

Sentiments expressed in Article 151 about production and the cultural economy are manifest in a range of subsequent policy statements, press releases, analyses and initiatives which are consolidated in the Green Paper of 2010. Subtitled *Unlocking the Potential of Cultural and Creative Industries*, this document was a direct response to an invitation from the European Parliament to create 'a genuine European strategy for culture' by addressing the ambiguities of Article 151 'to clarify what constitutes the European vision of culture, creativity and innovation' (European Parliament, 2008). The Green Paper thus repeats the validation of culture as a socially cohesive agent, opining that European audiences benefit from the circulation and consumption of works that offer them perspectives and challenges in order to 'understand and live in […] a more diverse cultural landscape' (European Commision, 2010, 4.2). Promoted beyond national boundaries, creative work is such that it aids European citizens 'to better know and understand each other's cultures, to appreciate the richness of cultural diversity and to see for themselves what they have in common'. (ibid.). The Green Paper also sought to 'elaborate political measures (…) in order to develop European creative industries' to address the failure of the Lisbon Agenda of 2000 that aimed to make the EU 'the most competitive and dynamic knowledge-based economy in the world capable of sustainable economic growth with more and better jobs and greater social cohesion' (European Council, 2000, 1.5).

So, while 'culture shapes our identities, aspirations and relations to others and the world' (European Commission, 2012), its grand promise is measured in more tangible manner for policymakers through its economic returns as a strategic fix for a set of general problems. Here, the most generous of assessments suggest that for the EU the cultural and creative sectors account for up to 4.5 per cent of GDP and nearly 4 per cent of employment: 8.5 million in directly related jobs although the impact of the sector has a role in underwriting many others. Thus, the creative sector is trumpeted for its impressive growth rate, where employment has proved to be resilient when compared to the EU economy as a whole, and 'In some cases, at local and regional level, strategic investments in these sectors have delivered spectacular results' (European Commission, 2013). In such rhetorical flourishes, cultural values and economic value often combine in the promise of creative industries for addressing the challenge of globalization and digitization, against which Europe must manage its heritage, identity and its economic competitiveness. Thus, whatever the generalizations of policy discourse and economic assessments, the primary output of creative industries is manifest in symbolic expression, meaning and experience. This quality is one that frames their discussion in a manner quite distinct from that of the utility of other areas of production.

There is a repeatedly chauvinistic quality to the discourse of cultural and creative industries that conflates a sense of native genius with anxieties about the fortunes of the economy. This is summarised by Xavier Greffe (2008: 163), for instance, who begins one survey 'Europe has long been regarded as the world's principle location of artistic creation and consumption, a centrality that is decreasing rapidly and worryingly'. Chauvinism is expressed elsewhere, direcetly, such as in the UK 'Britain is Great' international promotion campaign (UK Government, no date), and in French moves to protect the nation's film sector (Pratt, 2005: 42). There is a particular tension to be noted here, between the general and the particular. As Stuart Cunningham (2009: 378) suggests 'an increasing strain between aggressive market development strategies driving much else within the Union while cultural leaders champion exceptionalism'.

Here then, one detects something of the dichotomies and potential confusions structuring this area of policy formulation with which intermediaries and others involved in projects such as *Cross Innovation* have to contend. On one hand, culture as identity, drawing on a sense of diversity and heritage, is to be preserved and offers a bulwark against the threat of modernity. On the other, a particularly celebratory view of modernity is employed in promoting the value of creativity *per se*, and its potential for galvanizing economic growth with regards to digitization in particular. This evinces what Nicholas Garnham (2005: 22) has identified as 'the general Schumpeterian vision that now underpins much national and European Union economic policy under the "information society" label focused on innovation, innovation systems and national competition for the comparative advantage that successful innovation supposedly creates'. These qualities are particularly pronounced in ideas in the Green Paper about the potential transferability of creativity from cultural to other sectors. For instance, it is suggested that creative industries provide content that powers the take-up of digital devices and networks, contributing to the acceptance and further development of ICTs: 'As intensive users of technology, their demands also often spur adaptations and new developments of technology, providing innovation impulses to technology producers' (European Commision, 2010: 5).

Whatever the promise of the creative sector expressed in such instances, the policy ambition of nurturing production at a European level contends with a number of issues. First, despite the homogenizing pull exerted by the online world, there is considerable market fragmentation within the bounds of the EU, let alone in terms of export potential to the rest of the world. Production takes place amidst cultural traditions and expression from across the EU's 24 official languages, using three alphabets and around 60 regional and minority languages: 'This diversity is part of Europe's rich tapestry but it hinders efforts by authors to reach readers in other countries, for cinema or theatre goers to see foreign works, and for musicians to reach new listeners' (European Commission, 2013). Secondly, the EU's variety is one marked by uneven socio-economic development, exacerbated by the global financial crisis and widespread austerity measures that have been adopted as remedies for the financial crisis of 2008. As Flew notes, the crisis 'has exposed the highly indebted economies of Greece, Ireland, Portugal, Spain and Italy, who have been forced to undertake major public sector spending cuts in order to secure the euro as a common currency in the Eurozone region' (Flew, 2012: 34). Furthermore, and as Tony Judt (2010) has described in his reflection on the post-war fortunes of the continent, Europe can be thought of in terms of its geographical and cultural centre and periphery, bounded to the east by countries whose development has been hindered by decades within the Soviet zone of influence. Thirdly, is the sense that 'the arts' developed through a non-commercial logic. While this history might be over emphasized, the romantic insistence that 'culture begins where the market stops' (Greffe, 2008: 164) is a pervasive one, underwritten by a European tradition of public funding for the arts were the creative sector accommodates commercial and not-for-profit organisations (Huijgh, 2007).

On this last point, there is potential confusion – for policymakers, their agents, practitioners and scholars too – over the distinctions and relations of a public sector of activity in which cultural ideas appear to have primacy when compared to the return of commercial productions (Heartfield, 2005). This range and any distinctions are not made any clearer when operationalized by EU policy bodies that inform and authorise practice. In the *Cross Innovation* project for instance, the scope of the creative sector is one that includes: design; mobile and social media; games multimedia and web; architecture; performing and visual arts; audio-visual; music; open data; contemporary arts; publishing and fashion (the list of growth areas against which 'spillovers' are measured includes: healthcare; construction; catering; ICT; education; financial services and the maritime economy). In such instances it is useful to consider the creative sector as a continuum of interactions for instance via a metaphor of the ecosystem (as summarised by O'Brien 2013: 76). It is within the terms and parameters of this ecosystem and the cultures of creativity then that the role of the intermediary can be considered.

Cultural Intermediation as Process and Practice

The move from the cultural to the creative industries in international policy discourse has been widely analyzed. For critical commentators this entails more than a semantic shift, but reflects the reach of neo-liberalism into the aesthetic realm. Central to a critical view is the understanding of ideas and values associated with creativity that are diluted in contemporary cultural policy and across a wider vista. Chris Bilton (2011) has summarized some of the issues here, noting that the ubiquitous deployment of creativity as a totemic idea or contemporary buzzword entails an overuse that has emptied it of meaning. For Bilton, the use of 'creative' in conjunction with 'industries' or 'economy' marks a shift from the nuanced autonomy of the cultural 'towards a more individualized, commodified and managerial logic' (2011:1). In spite of its apparently unmanageable and magical qualities, creativity is enlisted in optimistic plans for economic growth of the kind itemized above, as well as the development of human potential, its logic serving contemporary notions of self-actualization, as traditional industries fall into decline in the developed world. As Jim McGuigan, one of the most critical commentators in policy debates, has written: 'In much cultural policy discourse and rhetoric today, the cultural has become an empty signifier that is articulated most typically for reductive economic purposes according to questionable politics' (McGuigan, 2004: 140). Thus, to paraphrase William Davies' recent work on neo-liberalism (2014), policy for the development and exploitation of creative industries might be understood as the *pursuit of the disenchantment of culture by economics*.

How then are we to understand the nature of this shift and, in light of the realm of EU governance and administration, how are the ambitious and often ambiguous terms of policy implemented and through which avenue and instruments? Justin O'Connor (2004) has argued for instance that the discourse of cultural industries is not simply one of making policy but can be considered as part of a wider set of issues of governance. The prodigious expansion of policy suggests a range of roles and practices at the interstices of their implementation and administration, between government bodies, producers and consumers, between the nature of economy and culture. For those involved – a 'species' of 'cultural intermediary' – this requires a concomitant 'set of self-understandings as part of the key skills in a new culture economy' (Ibid.: 40).

It is not necessary here to rehearse the reception of Pierre Bourdieu's (1984) identification of *new* cultural intermediaries as a particular group of commentators, entrepreneurs and taste-makers concerned with mass media. Suffice to say that there has been considerable dialogue on the extended meaning of the concept, its application and the identification of who exactly takes on this role (see, for instance: Hesmondhalgh 2006; Matthews and Smith, 2014; Warren and Jones, in press). As O'Connor (2015: 384) suggests, intermediaries are increasingly based in institutions:

> In designated creative industries offices, usually located in economic development sections of local authorities. They come from academics and managers in higher and further education who increasingly see the vocational implications of the creative economy as prime justification for the contemporary role of arts and humanities.

In the critique of contemporary policy discourse, it is hard not to conclude that this group are seen in generally disparaging terms although we would argue that there is little empirical engagement with the actual nature of intermediation as a constellation of processes and practices.

Taylor (2015: 364) argues that an understanding of intermediation is essential in 'explaining why the creative economy has become such a dominant and compelling (despite theoretical imprecision) strategic conceptualisation of the entwined economic, social and cultural dynamics of the regional spatial economy'. A concept of intermediation has certainly been valuable for understanding the shaping of the creative industries agenda and the role of policymakers and administrators therein and indeed the translation of ideas across a variety of contexts (e.g. Prince, 2010a; 2010b; 2012; Matthews and Smith, 2014). However, Taylor (2015) suggests that intermediation in the creative economy is marked by ambiguity and is thus relatively under-theorised. He offers a spatialised cultural political economy approach that addresses it in terms of its transactional, regulatory and strategic modalities. Here, the transactional involves the connections and relational activity of social and economic agents in the creative industries and policy realm. A modality of regulation encompasses the socio-political dimensions of transactions in terms of 'norms, rules and subjectivities'. At a strategic level, Taylor (2015: 364) identifies the enduring 'structural social forms through which these relations and subjectivities are reproduced'. Taking the interactivity between all three modalities is, for Taylor, a means of understanding the forms and efficacy of intermediation, of its structuring and legitimating role in embedding the creative economy as a subject that matters in the 'regional spatial order' (Ibid.).

That this economy matters in the spatial order of the EU can be appreciated in the detail that follows. The policy outlined above evinces strategies and norms around culture and its social and economic objectives. Its instruments mobilize a range of actors in projects such as *Cross Innovation*, which is assembled, managed and made meaningful from the transactions of intermediaries and cultural producers. In turn, the cultural perspectives, norms and dispositions – professional and personal – of those involved animate its processes and outcomes, the nuances and challenges of which are not always captured in the administrative assessment of milestones, key performance indicators and financial reports.

Cross Innovation: Policy and Practice and Process of Cultural Intermediation

In pursuit of the Lisbon Agenda, considerable material resources have been directed at nurturing economic growth through a variety of support mechanisms that are directly and indirectly aimed at the creative sector. *Creative Europe*, for instance, is the European Commission's Framework Programme supporting the cultural and media sectors with a budget of €1.46 billion (European Commission, no date). The *Cross Innovation* project, which aims to introduce the culture of the creative industries to other 'non-creative' sectors, is supported under the latest iteration of the broad interregional programme – Interreg IVC – itself financed through the European Regional Development Fund (ERDF). Over the last seven years Interreg has funded 14 projects in the creative sector specifically, involving 166 project partners from 85 regions and collecting 'Hundreds of individual examples […] of how a region or city has built on the experiences of their counterparts elsewhere to enhance their own policy and delivery strategies' (Interreg, *Creative Industries*, no date). The overall programme focuses on two thematic priorities. The first concerns innovation and the knowledge economy, incorporating amongst other things research and the development of technology. The second priority relates to environmental management, which incorporates cultural heritage. Interreg's objective is to improve the effectiveness of regional policies and development through the exchange of experiences among partners drawn from across EU member states. Its approach is predicated on a particular economic analysis that focuses on small and medium-sized enterprises (SMEs). While it is suggested that SMEs represent 99 per cent of all European business and create around 70 per cent of all jobs and GDP, the climate for such activities in Europe is seen as an unfavourable one and thus 'Common action is needed to reduce administrative burdens and increase the entrepreneurial spirit' (Interreg, *Thematic Programme Capitalisation*, no date).

The *Cross Innovation* project was built upon experience, insights and partnerships formed in two previous ventures funded under the Interreg scheme, and might therefore be considered something of a flagship project in terms of the durability of the ideas and practices it has pursued. These projects were: *Creative Metropoles* (2009–2011) and *Fostering Innovation in Economic Clusters of Cultural and Creative Enterprises* or *ECCE* (ecce-network.eu). *Creative Metropoles* was originated and led by cultural development professionals from the governments of various cities, a number of which form the current partnership. The ethos was prompted by contemporary ideas of how urban and regional development might be driven by creativity and culture evidenced in initiatives such as the annual award and recognition of a city as European Capital of Culture and a wide range of intermediaries (e.g. Griffiths *et al.*, 2003; O'Connor, 2004; Cooke and Lazzeretti, 2008; Van Winden, *et al.*, 2013). *ECCE* involved another consortium of European cities, their aim 'to support micro enterprises in the areas of art and culture in developing access to new markets beyond the conventional art and culture markets' (GründerZentrum, 2011: 2)

Lessons from these projects tended to confirm that quality of creative work identified by Richard Caves (2003: 74) reflecting on William Goldman that 'nobody knows' anything. While this is a rather hyperbolic statement (lots of people know a *lot* and invest a great deal in getting it 'right', albeit with no guarantee of *commercial* or *critical* success), it does point towards a range of challenges in the process of trying to maximize economic growth via the cultural sector, particularly through SMEs. As a report from *ECCE* noted of the heterogeneity of creative businesses – in terms of their organizational make-up and creative outputs – 'This leads to a still underdeveloped comprehensive industry awareness and little internal and external visibility of the overall industry and its relevance' (GründerZentrum, 2011: 6).

Such issues and the kinds of ambiguities about the creative sector that cultural intermediaries manage were apparent at the starting point for *Cross Innovation*. This involved scoping the understanding of ideas of project ideas in each locality. While many interlocutors consulted across the partnership 'were positive about the potential benefits of cross innovation and although able to cite anecdotal examples most acknowledged there is a lack of evidence about the impact' (Belgrave, 2012: 6). As was suggested in one report on the subject, Thomas Lämmer-Gamp (2014: 3) identified ideas of 'spillover effects' and 'cross-innovation' as popular policy terms in spite of the fact that 'it is not clear among the stakeholder [s] in the debate what is actually meant by the two terms'. Perceptions then tended to rehearse rather than challenge or investigate policy conclusions. In dealing with the particularly fissiparous characteristics of the creative sector and ambiguities about the core terms of its mission then, *Cross Innovation* as an intermediation process, can be understood in terms of the iterative approach taken towards disseminating a form of experiential 'cultural learning' (Kolb, 1984; Holden, 2008). This idea suggests how cultural means are used in order to aid individuals and communities of practice in this case to develop behaviours, skills, values or knowledge.

As Lämmer-Gamp puts it: 'Today's challenge for policy making is to create the mental basis for collaboration' (2014: 9). Thus, one mode of learning was directed at disseminating ideas about the norms of cultural industry practice – to creative workers as much as to those from without the sector and who might be engaged in cross innovation processes. In so doing then, intermediation on the scale of a project like this also offers a centripetal point for a form of management or even compensation, offering mechanisms of support and connectivity for creative industry SMEs which are, by their very nature, largely absent when compared with the economies of scale of major corporations. As Bastian Lange and Nina Lakeberg (2014) note in a report for *Cross Innovation,* the limited resources available to SMEs, whether financial or strategic in nature, mean that opportunities to network or experiment for instance become extremely valuable.

Building upon earlier projects, *Cross Innovation* objectives were collectively formulated and realized in a range of analyses and practical activities. These included: resourcing a study of cross innovation principles; developing a database cross innovation examples identified in each partner city; study visits for intermediaries and SMEs aimed at developing learning about innovation processes; the organization of 'policy clinics' for partner cities and local stakeholder groups; the development of a toolkit for galvanizing cross innovation practices and policies; delivery of workshops on internationalization and collaboration of SMEs across partner cities. Finally, each partner aimed to produce a bespoke local implementation plan for cross-innovation. The democratic and exploratory character of these mechanisms is expressed in the terms of a manifesto composed by Luca de Biase and Patrick van der Duin and published at the outset of the project. It commences with a statement that policymakers and their agents are tasked with dealing with changing social, cultural and economic demands but that 'Policy is not about dictating what is best for us; it should help us find our way to better modes of being, encouraging discovery, freedom and happiness' (cross-innovation.eu/practices/manifesto).

Consolidating the analyses and objectives of policy statements such as the EU Green Paper, and addressing obvious knowledge gaps, the project proceeded with a four-fold typology for locating cross innovation practice, anatomizing a wide set of normative practices associated with the cultural economy. These categories were then both investigative and suggestive, directing research and organizing understanding and reflections from partners when assessing local activities (Long, 2013). Over 40 organizations and their working practices were scrutinized in terms of the issues or problems each sought to address and the degree to which they addressed 'silo' or compartmentalized thinking and practice in nurturing 'spillovers' between sectors.

Of the four categories, *Smart Incentives* labels innovative modes of finance often involving a voucher scheme for in-kind support that is a means of enhancing the business development and acumen of SMEs as well as the skills of creative workers. A second framework for cross innovation practice is labelled *Spatial Cross-Collaboration*. This refers to the utility of co-working spaces such as 'incubators', 'fab-labs', science parks and cultural quarters. Earmarking spaces for creative workers and businesses is a form of clustering (Porter, 1990), reflecting the largely urban concentration of creative industries and the challenges of developing this work as part of the rural economy. As Lämmer-Gamp notes in recognizing the continued importance of geography in the digital age 'knowledge or technology are to a substantial degree local, meaning that so is spillover' (2014:40). Spaces are also important for the nurturing of sociality, a vital character for the development of SMEs in the creative sector and a characteristic that links them to ideas of the associational economy (Cooke and Morgan, 1999) and the nurturing of social capital (Olma, 2012). A third typology concerns *Culture-based Innovation*, which is used to refer to how the practices and working cultures of creative workers are introduced in to other sectors. As the EU *Green Paper* suggested 'it seems that creative innovation services provided by CCIs are inputs to innovative activities by other enterprises and organisations in the broader economy, thereby helping to address behavioural failures, such as risk aversion, status quo bias and myopia' (European Commission, 2010: 5). That there is some overlap between these frameworks for conceptualizing practices in the creative sector is manifest in the focus of the fourth label of *Brokerage*. In this project, *brokerage* names services and individuals – the *brokers* themselves – that build bridges between creative businesses and other industrial sectors.

The label of *broker* identifies the core role of particular individuals in making links between the creative sectors and other organizations. To some degree, the validation of the broker illustrated by the place of identifiable 'movers and shakers' in the sector reproduces an ideology of the unique artistic creative worker and transposes it to the broker in the role of intermediary facilitating innovation. The apparent indispensability of key individuals is bound up in their specialized understanding of the sector, and embodied expertise manifest in their diplomacy, dynamism, profile and contact list. As Prince (2014: 747) observes: 'Being an expert on matters of culture is popularly seen as a matter of subjective judgement. Cultural expertise rests on the ability to distinguish and valorise different cultural forms in a way that resonates with others possessing the same expertise, meaning expert judgements are as much of other people's judgements as of the forms in question'.

Of course, *brokerage* might also be taken to describe the very processes of *Cross Innovation* and cultural intermediation more generally. This is manifest in project strategies and events such as 'policy clinics' as well as surveys, presentation of 'best practice' online, and so on. These operate at the centre of a wide variety of empirical cultural work that evinces commercial as well as other imperatives. The kinds of projects that *Cross Innovation* has framed as exemplars of its ethos and for the satisfaction of policy objectives that include a range of interactions of creative sectors with wider industrial practices. These exemplify the promise of modernity expressed in EU policy for innovation 'the successful production, assimilation and exploitation of novelty in the economic and social spheres' (European Commission, 2003). Example projects include: the work of Pilzen's Lukáš Bellada, an expert in interactive surfaces, augmented reality and computer game development; *SmartGateCargo* a 'serious computer game' that aids the training of cargo managers at Amsterdam's Schiphol Airport; *Native Instruments*, a manufacturer of software and hardware for computer-based audio production and professional DJing based in Berlin; Rome's *EnLabs* is described as Italy's foremost 'open incubator and accelerator which also supports co-working' while Stockholm's *Kolonien* (*The Colony*) is a multidisciplinary co-working space and studio that develops companies, products and services through design, marketing, communication and concept development.

The degree to which collaboration and exchange has taken place in all cases assembled, however, is not always manifest in a manner liable to appear operationalized and transferrable to those looking for market optimization of creativity as 'industry'. Here, we see evidence of what David Hesmondhalgh (2007: 20–21) calls the 'commerce-creativity dialectic'. As the *Cross Innovation* manifesto states: 'Modern innovation processes resemble, on the one hand, a "division of labor" – a classical element of our free market society – but on the other, are rather responsive to societal demands and stress the need for cooperation between different societal actors. All this leads to responsible innovation that does not start with commercial interests'. Such statements point to a range of expectations and priorities, of differences and perspectives on culture and creativity that on this project reflect the diversity of the EU member states. This indicates the challenges for intermediaries in the drive for economic growth at the expense of the culture of the creative industries. As Lämmer-Gamp recognises:

> It is not only the potential clients in traditional industries have to shift their minds, a change of thinking is also required from the creative industry. Creative entrepreneurs are still very much focussed on social and cultural values. Often they equal thinking and acting economically with low quality. (Lämmer-Gamp, 2014:8)

'Social and cultural values' are certainly evinced in a variety projects such as the Birmingham-based *Digital Life Sciences* and the *Maverick Television Consortium* which seeks to improve user engagement with public health services. In the same city, Asian arts organisation *Sampad* works with technology providers to find new ways to bridge gaps between arts organisations and under engaged audiences. Pilsen's *Luckywaste* appropriates discarded material and 'upcycles' it into luxury products with a rational that stands in 'opposition to large, unindividualised companies, against the uniform fashion style of the young population, and helps emphasize the beauty of natural materials and carries an ecological aspect' (Quoted in Long, 2013: 15).

In the same vein we can point to projects such as the *Fluxus Ministry* of Lithuania. Inspired by the Fluxus Movement founded in the 1960s by Lithuanian-born American artist Jurgis (George) Mačiūnas, this privately funded initiative makes workspace available *gratis* and has gathered together 200 artists working in the old Soviet-era building of the former Ministry of Health in Vilnius and implemented a number of experimental art projects such as a 'Future City Lab'. Another privately funded project with public service goals is Vilnius's *Beepart* ('Be a part', 'bee art' or 'bee' in the sense of the 'common work' of the hive). It promotes cultural and social innovation, providing space for small community-based business initiatives, educational and experimental art projects. Run mainly by volunteers, its social role is apparent in its location in the Pilaitė district, one of the peripheral suburbs of the city. As outlined on its website, all entertainment and events in Vilnius, just as in other former Soviet-styled cities, are concentrated in the centre, usually the old town. This leaves the bigger part of the city – the so-called 'dormitory areas' – culturally bereft. Addressing this issue, *Beepart* serves as a meeting point for local community members and artists, aiming to be a lab for any experimental idea that could help improve the social and cultural environment of the community and, open to community members as well as encouraging participants to take an active role in cultural and social projects.

While commercial projects abound in the repertoire of case-studies of *Cross Innovation* from Estonia and Lithuania, the examples of *Fluxus* and *Beepart* are useful for illustrating how issues of cultural difference become more visible when one considers member states that were until relatively recently within the sphere of the Eastern Bloc. For instance, Liutkus (2014: 2) reports on the 'heated debates' around cultural policy that followed Lithuania's own version of 'Perestojka'. Politicians, artists, philosophers and administrators addressed issues of cultural democratisation, the protection of cultural heritage and how to give new guarantees for the freedom and diversity of creative activity as well as cultural self-governance, bound up in a need for the decentralization of authority. Kirill Razlogov (2008) has reflected on the difficulties facing the cultural and creative sector in post-socialist states: state-protectionism gone, and with them the instruments of support and an atrophied sense of cultural enterprise born of system of 'equality for both the hard-working and the lazy, the talented and the hacks' (Ibid.: 173). The new conditions are captured in a description from one film director who suggested that in the past, while living a cage and under surveillance, artists were well fed and protected: 'Now we find ourselves in the jungle, free but fighting for food and having the choice between kill or get killed' (Alexei German, quoted in Razlogov, 2008: 173). On Estonia, Lassur et al. (2010) note how the Ministry of Culture adapted the British definition of creative industries, inserting into its widely circulated celebration of individual origination a qualifying clause about 'collective creativity'. Signing up for EU projects was also sometimes born of pragmatism resulting from the lack of available funding for the cultural sector.

Rozlogov suggests of the 'pathologies of transition' that there is a consensus about the difficulty of adapting traditional 'Eastern' values to Western practices. 'The mis-adaptation or radical refusal of modernization demonstrates the impasse' (Razlogov, 2008: 176). Indeed, if modernization is represented in policies for exploiting culture in the form of the creative industries, it has to work with the 'defensive strategy' identified in Mikko Lagerspetz and Margaret Tali's (2014: 4) study of Estonia, as well as observations about the less than universal nature of creative industries (Rozentale and Lavanga, 2014). In Estonia 'the cultural workers themselves feel they must jointly defend themselves against the invasion of mass culture, against the insecurity created by a dependence on market mechanisms' (Lagerspetz and Tali, 2014: 4). As one of the operatives of Lithunia's *Beepart* commented when describing their activity and aims as more 'cultural' than commercial: 'So I don't know if we can call this an industry' (Andrius Ciplijauskas, email to Paul Long, 20 August 2014).

Conclusions

There is a wide literature that explores the forms and efficacy of cultural policy outcomes from the fields of business and economics and which can be highly critical of the broad return and value of EU directives (see, for instance Cooke, and De Propris, 2012). Perhaps local projects like this example then need more scrutiny for how to make sense of their value on these terms. Perhaps they need closer attention from critical theorists too, in order at least to address the empirical gaps in their knowledge, as if motivation, practice and social relations can simply be inferred from a starting point in a treaty or paper (For a critique see: O'Brien and Miles, 2010).

The policy injunctions of the EU then, transmitted via instruments such as Interreg and manifest in projects such as *Cross Innovation,* present a rich source for empirical consideration of cultural intermediation. As we hope we have shown, this detail offers an understanding of intermediation as process, a translation 'between the language of policymakers and that of the cultural producers' (O'Connor, 2004: 40), between economic imperatives and the less instrumental cultural objectives of the culture industries. It illuminates the interplay of transactions, strategies and regulations, of the everyday norms that make the creative economy a reality.

Mindful of our own roles and interest in this and other projects, it would be disingenuous to portray this project as simply an unreflexive and banal administrative driver of policy, or a site of autonomous cultural practice outside of demands for economic development. Well, we would say that wouldn't we? But we do welcome the scrutiny of researchers, questions and frameworks that lead us beyond the satisfaction of the kinds of milestone reports and publicity releases which satisfy funders that projects are 'on target'. Herein, it is people and social and cultural relations that matter, in working through policy in ways that are not always easily captured. Intermediation happens as a process of negotiation, of encounter, problematization, and perhaps compromise in ways deserving of further scrutiny.

As we have suggested, the nature of the creative industries as symbol producers is what gives them their particular character. They embody aspects of the lived and imaginative culture of member states, regions and ethnicities in particular ways and have a part to play in the politics and identity of the EU project. Such projects are framed as part of a long-term intervention that looks ahead to 2020. Cultural policy therefore works on a timeframe often unavailable or unattractive to locally elected politicians working on a relatively short time in office. Cultural policy may indeed be read as an 'empty signifier' but for intermediaries and other agents in *Cross Innovation* and projects like it, a complex site is authorising for the making of the European project and what democracy and citizenship might mean.

References

Bakhshi, H. and Throsby, D. (2010) *Culture of Innovation: An economic analysis of innovation in arts and cultural organisations*. London: Nesta.

Bakhshi, H. and Throsby, D. (2012) New technologies in cultural institutions: theory, evidence and policy implications. *International Journal of Cultural Policy* 18(2), pp. 205–222.

Banus, E. (2002) Cultural policy in the European Union and the European identity, in M. Farrell, Mary, S. Fella, and M. Newman (eds) *European Integration in the Twenty-first Century: Unity in diversity?* London: Sage, pp. 158–183.

Barnett, C. (2001) Culture, policy, and subsidiarity in the European Union: from symbolic identity to the governmentalisation of culture. *Political Geography* 20(4), pp. 405–426.

Belgrave, J. (2012) *Cross Innovation Partnership: short study on cross innovation* <http://www.cross-innovation.eu/wp-content/uploads/2013/05/Cross-Innovation-Short-Study-final.pdf> Accessed 1 June 2014.

Bilton, C. (2011) *Creativity and Cultural Policy*. London: Routledge.

Bourdieu, P. (1984) *Distinction*. Oxford: Routledge.

Caves, R.E. (2003) Contracts between art and commerce. *Journal of Economic Perspectives* 17(2), pp. 73–84.

Cooke, P. and Morgan, K. (1999) *The Associational Economy: firms, regions and innovations*. Oxford: Oxford University Press.

Cooke, P. and Lazzeretti, L. eds (2008) *Creative Cities, Cultural Clusters and Local Economic Development*. Cheltenham: Edward Elgar Publishing.

Cooke, P. and De Propris, L. (2012) For a resilient, sustainable and creative European economy, in what ways is the EU important? in P. Cooke, M. Parrilli and J. Curbelo, (eds) *Innovation, Global Change and Territorial Resilience*. Cheltenham: Edward Elgar Publishing, pp. 403–428.

Cunningham, S. 2009. Trojan horse or Rorschach blot? Creative industries discourse around the world. *International Journal of Cultural Policy* 15(4), pp. 375–386.

Davies, W. (2014) *The Limits of Neoliberalism: authority, sovereignty and the logic of competition*. London: Sage.

European Commission (No date) Creative Europe *What's it about?* <http://ec.europa.eu/programmes/creative-europe/> Accessed 1 September 2014.

—— 13 May 2003 *Innovation and the Lisbon strategy* <europa.eu/legislation_summaries/other/n26021_en.htm> Accessed 1 November 2014.

—— 27 April 2010 *COM (2010) 183 Final GREEN PAPER: Unlocking the potential of cultural and creative industries* <http://eur-lex.europa.eu/legal-content/EN/TXT/HTML/?uri=CELEX:52010DC0183&from=EN> Accessed 1 December 2014.

—— 26 September 2012 *Communication from the Commission to the European Parliament, The Council, The European Economic and Social Committee and The Committee of the Regions: Promoting cultural and creative sectors for growth and jobs in the EU* <http://eur-lex.europa.eu/legal-content/EN/ALL/;ELX_SESSIONID=tGDMTtwDnnVtMzvDfyQmmVGqKHpn199DTjcbYwPVhgjKh42v0WBp!-15732272?uri=CELEX:52012DC0537> Accessed 15 August 2014.

—— 19 November 2013 *Memo. Creative Europe: frequently asked questions* Brussels/Strasbourg <http://europa.eu/rapid/press-release_MEMO-13–1009_en.htm?locale=FR.> Accessed 1 December 2014.

European Community (1997) *Treaty Establishing the European Community (Amsterdam consolidated version) – Part Three: community policies – Title XII: Culture – Article 151 – Article 128 – EC Treaty (Maastricht consolidated version) – Article 128 – EEC Treaty* <http://eur-lex.europa.eu/legal-content/EN/TXT/HTML/?uri=CELEX:11997E151&from=EN> Accessed 1 November 2014.

European Council (2000) *Presidency Conclusions, Lisbon European Council, 23 and 24 March 2000.*<http://www.consilium.europa.eu/uedocs/cms_data/docs/pressdata/en/ec/00100-r1.en0.htm> Accessed 1 December 2014.

European Parliament (2008) *Resolution of 10 April 2008 on Cultural Industries in the Context of the Lisbon Strategy* <http://www.europarl.europa.eu/sides/getDoc.do?type=TA&reference=P6-TA-2008–0123&language=EN&ring=A6–2008–0063> Accessed 1 December 2014.

Flew, T. (2012) *The Creative Industries: culture and policy* London: Sage.

Garnham, N. (2005) From cultural to creative industries: an analysis of the implications of the "creative industries" approach to arts and media policy making in the United Kingdom. *International Journal of Cultural Policy* 11(1), pp. 15–29.

Gordon, C. (2010) Great expectations–the European Union and cultural policy: fact or fiction? *International Journal of Cultural Policy* 16(2), pp. 101–120.

Greffe, X. (2008) European cultural systems in turmoil, in H.K. Anheier and Y.R. Isar (eds) *Cultures and Globalization: The cultural economy.* London: Sage, pp. 163–171.

Griffiths, R., Bassett, K., and Smith, I. (2003) Capitalising on culture: cities and the changing landscape of cultural policy. *Policy & Politics* 31(2), pp. 153–169.

GründerZentrum Kulturwirtschaft (2011) *ECCE Innovation Summary on Public Procurement* Aachen <http://ecce-network.eu/rtefiles/File/report-on-public-procurement> Accessed 1 December 2014.

Heartfield, J. (2005) *The Creativity Gap*. Blueprint. <www.wdis.co.uk/blueprint/broadsides.asp> Accessed 1 December 2014.

Hesmondhalgh, D. (2006) Bourdieu, the media and cultural production. *Media, Culture & Society* 28(2), pp. 211–231.

Hesmondhalgh, D (2007) *The Cultural Industries*. London: Sage, pp. 20–21.

Holden, J. (2008) *Culture and Learning: towards a new agenda*. London: DEMOS.

Huijgh, E. (2007) Diversity united? Towards a European cultural industries policy. *Policy Studies* 28(3), pp. 209–224.

INTERREG IVC *Creative Industries*. No date. <http://www.interreg4c.eu/creativeindustries/> Accessed 1 October 2014.

— Thematic Programme Capitalisation. No date. <http://www.interreg4c.eu/capitalisation> Accessed 1 October 2014.

Judt, T. (2010) *Postwar: a history of Europe since 1945*. London: Random House.

Kolb, D. (1984) *Experiential Learning: experience as the source of learning and development* Vol. 1. Englewood Cliffs, NJ: Prentice-Hall.

Lagerspetz, M. and Tali, M. (2014) Country Profile: Estonia *Council of Europe/ERICarts: Compendium of Cultural Policies and Trends in Europe, 15th edition 2012.* Updated 2014. <http://www.culturalpolicies.net/down/estonia_092014.pdf> Accessed 1 August 2014.

Lämmer-Gamp, T. (2014) *Creative Industries: policy recommendations – promotion of cross-innovation from creative industries*. Berlin: VDI/VDE Innovation + Technik GmbH.

Lange, B., and Lakeberg, N. (2014) *Feasibility Study on SME Internationalisation: a report*. Berlin: Multiplicities/Cross-Innovation.

Lassur, S., Tafel-Viia, K., Summatavet, K. and Terk, E. (2010) Intertwining of drivers in formation of new policy focus: case of creative industries in Tallinn. *Nordic Journal of Cultural Policy* 1(13), pp. 59–86.

Liutkus, V. (2014) Country Profile: Lithuania. *Council of Europe/ERICarts: Compendium of Cultural Policies and Trends in Europe, 14th edition 2012* <http://www.culturalpolicies.net/down/lithuania_102014.pdf> Accessed 1 August 2014.

Long, P. (2013) *Cross Innovation: a report on local best practice* <www.cross-innovation.eu/wp-content/uploads/2013/05/Case-Study-Review-For-Publication.pdf> Accessed 1 June 2014.

Matthews, J. and Smith, J., eds. (2014) *The Cultural Intermediaries Reader*. London: Sage.

McGuigan, J. (2004) *Rethinking Cultural Policy*. Maidenhead: Open University Press.

O'Brien, D. and Miles, S. (2010) Cultural policy as rhetoric and reality: a comparative analysis of policy making in the peripheral north of England. *Cultural Trends* 19(1–2), pp. 3–13.

O'Brien, D. (2013) *Cultural Policy: management, value and modernity in the creative industries* London: Routledge.

O'Connor, J. (2004) Cities, culture and "transitional economies", in D. Power and A. Scott (eds) *Cultural Industries and the Production of Culture*. London: Routledge, pp. 37–53.

O'Connor, J. (2015) Intermediaries and imaginaries in the cultural and creative industries. *Regional Studies* 49(3), pp. 374–387.

Olma, S. (2012) *The Serendipity Machine: a disruptive business model for society 3.0*. Amersfoort: Lindonk & De Bres.

Porter, M. (1990) *The Competitive Advantage of Nations*. New York: The Free Press.

Pratt, A. C (2005) Cultural industries and public policy: an oxymoron? *International Journal of Cultural Policy* 11(1), pp. 31–44.

Prince, R. (2010a) Fleshing out expertise: the making of creative industries experts in the United Kingdom. *Geoforum* 41(6), pp. 875–884.

Prince, R. (2010b) Globalizing the creative industries concept: travelling policy and transnational policy communities. *The Journal of Arts Management, Law, and Society* 40(2), pp. 119–139.

Prince, R. (2012) Policy transfer, consultants and the geographies of governance. *Progress in Human Geography* 36(2), pp. 188–203.

Prince, R. (2014) Calculative cultural expertise? consultants and politics in the UK cultural sector. *Sociology* 48(4), pp. 747–762.

Psychogiopoulou, E. (2008) *Integration of Cultural Considerations in European Union Law and Policies*. Leiden: Martinus Nijhoff Publishers.

Razlogov, K. (2008) Countries in transition: which way to go? in H.K. Anheier and Y.R. Isar (eds) *Cultures and Globalization: the cultural economy*. London: Sage.

Rozentale, I., and Lavanga, M. (2014) The "universal" characteristics of creative industries revisited: the case of Riga City. *Culture and Society* 5, pp. 55–64

Sassatelli, M. (2002) Imagined Europe. *European Journal of Social Theory* 5(4), pp. 435–451.

Taylor, C. (2015) Between culture, policy and industry: modalities of intermediation in the creative economy. *Regional Studies* 49(3), pp. 362–373.

Tzaliki, L. (2007) The Construction of European Identity and Citizenship through Cultural Policy, in K. Sarikakis (ed.) *Media and Cultural Policy in the European Union. European Studies: A Journal of European Culture, History and Politics* 24, pp. 157–182.

UK Government (no date) *Britain is Great* <www.gov.uk/britainisgreat> Accessed 1 June 2014.

Van Winden, W., De Carvalho, L., van Tuijl, E., van Haaren, J., and Van den Berg, L. (2013) *Creating Knowledge Locations in Cities: innovation and integration challenges* Vol. 54. London: Routledge.

Warren, S. and Jones, P. (in press) Local governance, disadvantaged communities and cultural intermediation in the creative urban economy. *Environment and Planning C.*

Chapter 10
Conclusion: The Place of Creative Policy?

Phil Jones and Saskia Warren

There is no singular creative *economy*, but a number of creative *economies* encompassing everyone from the street artist, to the opera singer, the web designer to the community project worker. Beyond romantic notions about the creative impulse, perhaps the strongest bond between these different subsectors is a nested set of policy frameworks cutting across local, regional, national and supranational scales. Within this policy discourse, creative activity is seen as having the potential to deliver economic and social benefits. The extent to which these benefits are as great as their promoters claim is, of course open to question. Following Thornham (2014) one can see a division between creative *products* made by individuals and creativity as *process*, a mode of engagement aimed at groups of people. The celebration of the individual creative producer fits neatly within a neoliberal discourse of entrepreneurialism. This sits rather uncomfortably with the idea of creative processes allowing 'communities' (hazily defined) to experiment with new practices and learn new ways to express themselves. A process-led framing of creativity is often explicitly situated within a social justice agenda that works with deprived communities, positioned against a retreat of state responsibility for creating a more equal society.

Within policy discourses, 'communities' are asked to perform a significant role, mitigating the inequalities that are driven by this celebration of the individual and the undermining of the state as an agent of social justice (Imrie and Raco, 2003). In practice, of course, communities are every bit as fluid and contingent as the creative sector itself. Policy, economy and community do, however, have a lived reality, brought into being at specific times and particular *places*. One of the key themes uniting the contributions to this volume is in considering how place shapes and is shaped by the ways in which creative economies and communities intersect. As a concept, place is not unproblematic, however. The trap to avoid whenever discussing place is in seeing it as something which is simply related to territory – a unit of so and so square kilometres, with such and such boundaries. Ash Amin (2004) cautions us against such an approach as it ignores the ways that places are relational, constructed at different scales and at different times by different people. By trying to fix place to specific territories – and Amin concentrates on the regional scale – we run the risk of creating a somewhat inward looking understanding of the world. In the context of creative economies and communities the temptation when talking about the importance of place is to juxtapose the local as grassroots, authentic, resistant against the global as corporate, fake and exploitative. This kind of binary is unhelpful, first, in that it fails to acknowledge how place can cut across the global and local scales. Secondly, the local-good, global-bad binary fails to challenge broader structural inequalities while at the same time idealising 'communities' as the locus of meaningful human interaction. The locally-grounded and territorially-specific can be highly conservative and inward looking just as global approaches can drive innovation and social justice. Understanding how place is constructed at different scales across different communities allows us to understand the messy ways in which creative economies are enacted.

In Daya's chapter we are reminded not to simply celebrate grassroots creative production either as a model of neoliberal entrepreneurialism or for its power to assert place-based cultural identity; this would be to gloss over the desperate income inequalities driving informal creative producers' reliance on beadwork to scrape a living. Place shapes the work, but these producers no more live in a singular southern African 'place' than they constitute a singular community of artists. This becomes particularly complicated where individuals have migrated from elsewhere and construct place across national boundaries. Those who are artistically gifted / entrepreneurial and whose residency is uncontested will find themselves in a relatively privileged place. Those struggling with lesser talent or problematic citizenship status will be pushed to the socio-economic margins.

A major critique of the Florida and Landry-inspired policy prescription advocating creativity as a universal good is that it ignores the specific challenges facing particular places. The recipe of now somewhat standardised cultural responses – a new art gallery, co-working space for creative microbusinesses, a raft of community arts projects – will not solve every problem, everywhere. In fairness to Florida, he never claimed that such things would. It is clear that asking culture to solve structural economic problems facing places and communities seems optimistic at best, duplicitous tokenism at worse. It is in this context that Layard et al.'s contribution is so important, highlighting the varied approaches to the law that community organisations use to manage the challenges they face, even where they fall under what is nominally the exact same legal framework. Even the law, it seems, is not as absolute as it first appears, giving scope for people to redefine what is normal (and legally acceptable) in different places. The graffiti in Stokes Croft thus stops being anti-social and, reimagined as street art, becomes a key element of place identity (and, incidentally, place marketing). We can see a difference here in the ways that creative *production* (in the case of the graffiti) and creative *process* (in the case of Northern Youth's engagement with the Community X-Change project) have managed their relationships legally. For Northern Youth, there is a clear frustration that the creative process did not create a space to tackle the social justice issues they were engaging with. Nonetheless, the Stokes Croft case is problematic in that by enacting a highly *visible* creative practice, one particularly active group of people in a neighbourhood has had their vision of place given official (legal) sanction, imposing this on others living there who do not necessarily share that view of what the place should be.

In broader cultural policy we can see a division between initiatives that are part of city reimaging exercises and those that intervene more directly at the neighbourhood scale. New art galleries, waterfront redevelopments, city centre festivals and so on may have important social outreach elements to their work, but primarily they are conceived to put the host town 'on the map' – big bang attempts to attract external visitors and investment. Many of the projects discussed in this volume exist on the periphery of this mainstream activity and are much less *visible* except where people choose to engage with them. To many people living in and passing through the Balsall Heath area of Birmingham the creative activity happening in the Old Printworks (OPW) – home of the *Some Cities* project discussed by O'Brien – is simply invisible. This is what makes the idea of participatory evaluation so challenging. If we have a policy discourse that culture is good for people then it follows that any public money spent on cultural activity should target as many people as possible in order to spread the benefits of that spend as widely as possible. Should projects like *Some Cities* be seen as a failure if fewer than x-thousand people engage with them because of their peripheral position in the mainstream cultural landscape and limited visibility in the physical landscape? Or should we look for quality of engagement and for measures of substantive (positive) changes to people's lives? If we consider that individuals and communities will benefit from an activity in different ways, then the idea of asking the people who actually want to take part in a project to define the measures of success makes sense.

We divided this collection into two sections, examining first the ways that creative practice (both as production and process) has been tied to community-making and second how policy frameworks relate to these different kinds of creative practices. While this can undoubtedly be seen as a somewhat arbitrary division, we use it to indicate that these three elements – policy, creativity and community – are not always as well connected as they might be. In the UK, the Arts and Humanities Research Council has supported a good deal of work recently exploring these disconnects as part of its *Connected Communities* agenda – indeed, several contributors to this volume have received funding through this scheme. Part of *Connected Communities*' purpose was to examine how policy, arts practice and communities intersect, resulting in a great many richly detailed case studies. Examining these case studies makes it clear, however, that broadly similar UK policy frameworks aimed at engaging communities with creative practice play out in distinctive ways across different locales. This underlines the value of Wollaston and Collins' chapter in giving insights into how policymakers can adapt a national framework to attempt something more innovative at the local scale that takes place-specificity seriously.

Long and Harding's contribution and that by Acott and Urquhart take a wider view by examining pan-European examples. The photographic explorations of coastal fishing settlements resist attempts to flatten these into a singular experience of place, evoking common atmospheres whilst emphasising unique local identities. The scale at which policy operates is significant here. This has always been particularly important in European policymaking where the principle of subsidiarity emphasises the need not to overwhelm the regional and particular with the pursuit of a supranational framework; instead decisions should be taken at the most appropriate scale for the types of policy being enacted. Notions of place, culture and community can also be considered to operate across multiple scales. The idea of Europe having a place identity only emerges in juxtaposition to the global scale – Europe as a different *place* to Africa for example. Below that scale a multiplicity of contested national, regional and local places emerge, depending on the issues being considered. The cross-innovation project discussed by Long and Harding brings these issues of scale to the fore by emphasising the need to maximise the benefit of interactions between creative economies and other sectors across the EU. On one level cross-innovation can be read as a centrist project, attempting to encourage conformity to a supranational policy framework. This occurs, however, within the context of EU Article 151 walking a fine line in between the aspiration to reduce conflict by creating a shared cultural identity and acknowledging the importance of not destroying the diverse cultural heritages that exist across the continent.

Both these examples bring us back to the point about disconnects, or, perhaps, weak ties, between creative practice, policy and ideas of community. The cross-innovation project obfuscates the ways in which a culturally-driven economic policy tries to enact a particular kind of pan-European *place*. This place brings together a community across Europe that is excited by the potential of the digital economy for creative producers, but does not necessarily ground that community of digital enthusiasts in the challenges facing specific national, regional and local places. Similarly, changes to supranational scale regulation on fisheries and wider economic pressures have impacted the coastal towns of north-west Europe in different ways. While the policy frame seeks to protect a particular environment and ecosystem in this part of the world, the varied cultural landscapes related to these eco-systems mean that there is no single sense of place cutting across these communities. In both cases, policy as it relates to creative economies is not always as well connected as it could be to issues of socio-economic justice facing communities across a number of scales.

One of the criticisms of the creative economy discourse is its urban focus. Urban and rural are interconnected in a variety of ways and should not be considered as completely independent of each other, but we should be mindful of the fact that the experience of place outside towns and cities remains distinct. Thus the issues of connectivity raised by Roberts and Townsend echo some of the wider policy concerns about digital 'not-spots' hampering the development of the rural economy. There has long been a strong 'escape the cities' strand to artistic production, from the expressionists retreating to the Moritzburg Lakes in the years prior to World War I, to the American land artists attracted to the wide open spaces of the remote western states in the 1950s and 1960s. The emphasis in creative economy discourses of ever faster connection in order to monetise the products generated by artists and creative producers does not always sit comfortably with those individuals seeking inspiration in remoteness and disconnection.

One should not, however, treat the rural simply as a place of inspiration for tortured artists tired of life in the city. There are many reasons why people still live outside the major cities and this raises broader questions about how the claimed benefits of the creative economies can be distributed beyond its metropolitan heartlands. We see this question within the discussions of changing museum and library functions in Poland's core and peripheral areas discussed by Działek and Murzyn-Kupisz. There is great potential for museums and libraries to move beyond their original remits and use their resources to try to find ways of fostering community cohesion. Adapting to new roles can be difficult for institutions, however, particularly when resource tends to be concentrated in the elite facilities of the major cities. This is a familiar problem in the UK where public arts spending is disproportionately concentrated in the major institutions based in London despite the capital arguably needing less public assistance to pump prime its creative economies.

Where we get to by the end of this volume is to ask questions about whether creative economies can meet the myriad and often contradictory expectations with which the sector has been burdened. There is no doubt that creative practice can be highly valuable in city marketing, digital innovation, driving economic growth and that engagements with creative processes can contribute to fostering wellbeing, raising aspirations and increasing social cohesion. But there is no way that creative economies can and should be expected to solve a set of intractable socio-economic problems alone. In attempting to think through the role of place within creative economies and community discourses one of the major contributions made by this collection is in thinking about the reterritorialization of creative economies. Much of the work on creative economy is implicitly inspired by Porter's (1998) work on clustering. Porter argued that in order to drive innovation, businesses working in similar sectors cluster in the same geographic location – Silicon Valley being the classic example – to exchange tacit knowledge and share a talent pool. The idea that businesses need to cluster by sector in order to succeed means that thinking about the creative economies has a distinctly metropolitan tinge. In this volume contributors have explored slums, seascapes, the rural and smaller towns as well as the big cities to think through spaces that are physically peripheral and practices that have been made peripheral to the mainstream ('below' the law, beyond the pale!). More than this, as well as expanding the territorial reach of creative economies, we have examined how place, multiscalar and relational, is key to mediating the relationship between creativity, policy and community.

Place-making is based on both collaboration and conflict (Pierce et al. 2011) and this goes to the heart of what we have sought to achieve with this book. Places are where creative economies, policy and community are created, enacted, intersect and frequently clash. Creative economies cannot be the solution to all of society's ills, but by examining where and how creative economies are practiced, engaged with and evaluated, we start to indicate the direction for policymakers, researchers, practitioners and residents seeking to enhance socio-economic, environmental, educational and cultural outcomes for communities.

References

Amin, A. (2004) Regions unbound: towards a new politics of place *Geografiska Annaler: Series B, Human Geography* 86:1, 33–44.

Imrie, R. and Raco, M. (2003) Community and the changing nature of urban policy, in Imrie, R. and Raco, M. (eds) *Urban Renaissance? New Labour, Community and Urban Policy*, Policy Press: Bristol, 3–36.

Pierce, J., Martin, D. and Murphy, J. (2011) Relational place-making: the networked politics of place *Transactions of the Institute of British Geographers* 36:1, 54–70.

Porter, M. (1998) Clusters and the new economics of competition *Harvard Business Review* 76:6, 77–90.

Thornham, H. (2014) Claiming 'creativity': discourse, 'doctrine' or participatory practice? *International Journal of Cultural Policy* 20:5, 536–52.

Index